HIGHER
GCSE Mathematics for Edexcel

HOMEWORK BOOK

JEAN LINSKY

SERIES EDITOR: ALAN SMITH

WITHDRAWN

HODDER
EDUCATION
PART OF HACHETTE LIVRE UK

D1324686

This high quality material is endorsed by Edexcel and has been through a rigorous quality assurance programme to ensure that it is a suitable companion to the specification for both learners and teachers. This does not mean that its contents will be used verbatim when setting examinations, nor is it to be read as being the official specification – a copy of which is available at www.edexcel.org.uk

Orders: please contact Bookpoint Ltd, 130 Milton Park, Abingdon, Oxon OX14 4SB.
Telephone: (44) 01235 827720. Fax: (44) 01235 400454. Lines are open 9 am to 5 pm, Monday to Saturday, with a 24-hour message answering service.
Visit our website at www.hoddereducation.co.uk

© Jean Linsky and Alan Smith 2006
First published in 2006 by
Hodder Education, part of Hachette Livre UK
338 Euston Road
London NW1 3BH

Personal Tutor CD: © Sophie Goldie, Alan Smith, 2006, with contributions from
Andy Sturman; developed by Infuze Limited; cast: Nicolette Landau; recorded at Alchemy Soho.

Impression number   10 9 8 7 6 5 4 3
Year   2011   2010   2009   2008

Cover illustration © David Angel @ Début Art
Illustrations © Barking Dog Art
Typeset in Great Britain by Tech-Set Ltd
Printed and bound in Great Britain by Martins The Printers, Berwick-upon-Tweed.

A catalogue record for this title is available from the British Library.

ISBN: 978 0340 913 628

Contents

Introduction

This Homework Book has been written to support the Hodder Murray Student's Book for the two-tier Edexcel GCSE Mathematics (Higher Tier) specification for first teaching from September 2006. The book has been written with reference to the linear specification, though it can also be used effectively to cover the required content for the modular course.

The sequence of the material mirrors that of the Student's Book, so that, for example, Exercise 7.2 of the Homework Book provides further practice questions matching the content of Exercise 7.2 in the Student's Book.

The first 23 chapters target the material that is typically needed to reach a grade B, while chapters 24 to 32 are essentially A and A* grade.

An approximate indication of the level of difficulty of each exercise is given by the following set of icons:

no icon   Low
■   Medium
▲   High
★   Extension

More practice helps to reinforce the ideas you have learned and makes it easier to remember at a later stage. If, however, you do forget, further help is at hand. As well as the Student's Book, with this book there is also a Personal Tutor CD-ROM. This contains worked examples on key topics to help you recall what you may have forgotten.

Look out for this symbol **pt** next to an exercise; it tells you that there is a relevant worked example on the CD-ROM.

For all the questions in this book, an icon shows you whether calculators are permitted or not, and you should follow this advice carefully in order to build up the necessary mixture of calculator and non-calculator skills before the examination.

All of the content of this book has been checked very carefully against the new GCSE specification, to ensure that all examination topics are suitably covered. In addition, practice exam-style papers, with full mark schemes, are available in Hodder Murray's accompanying Assessment Pack.

You will find the answers to this homework book in the Higher Assessment Pack.

CHAPTER 1

Working with whole numbers

EXERCISE 1.1

Work out the answers to these problems in your head.

1 38 + 27 + 594 + 12   **2** 234 + 356 + 41   **3** 55 + 739 + 184

4 743 + 136 + 87   **5** 32 + 33 + 34 + 35 + 36   **6** 26 + 53 + 17 + 49

7 38 + 49 + 795   **8** 437 − 99   **9** 138 + 381

10 573 − 399

Use any written method to work out the answers to these problems. Show your working clearly.

11 825 + 67   **12** 170 + 376   **13** 826 − 394

14 890 − 368   **15** 2647 + 3691   **16** 509 + 417 + 32

17 7302 − 4798   **18** 7007 − 3678

19 An aircraft can carry 326 passengers when all the seats are full, but today 78 of the seats are empty. How many passengers are on the aircraft today?

20 The attendances at a cinema complex were 379 (Monday), 526 (Tuesday) and 804 (Wednesday). How many people attended in total?

EXERCISE 1.2

Use short multiplication to work out the answers to these calculations.

1 235 × 5   **2** 482 × 7   **3** 384 × 3   **4** 168 × 9

5 514 × 6   **6** 690 × 7   **7** 723 × 8   **8** 888 × 4

Use any written method to work out the answers to these problems. Show your working clearly.

9 573 × 21   **10** 389 × 32   **11** 704 × 45   **12** 616 × 82

13 461 × 63   **14** 902 × 56   **15** 157 × 97   **16** 833 × 74

17 A company has 36 coaches and each coach can carry 43 passengers. How many passengers in total can all the coaches carry?

18 I have a set of 24 encyclopaedias. Each one has 288 pages.
How many pages are there in the whole set?

19 Jon buys 18 stamps at 26 pence each and 14 stamps at 47 pence each.
How much does she spend in total?

20 A cinema has 43 rows with 26 seats in each row.
Calculate the total number of seats in the cinema.

EXERCISE 1.3

Use short division to work out the answers to these calculations.
(Three of them should leave remainders.)

1 812 ÷ 6   2 1364 ÷ 4   3 9279 ÷ 9   4 6832 ÷ 7

5 6895 ÷ 5   6 3456 ÷ 3   7 2394 ÷ 8   8 500 ÷ 6

Use long division to work out the answers to these problems. Show your working clearly.
(Only the last two should leave remainders.)

9 1323 ÷ 21   10 1984 ÷ 16   11 9750 ÷ 25   12 7776 ÷ 32

13 6372 ÷ 18   14 9196 ÷ 22   15 5367 ÷ 31   16 4055 ÷ 27

17 £1164 is shared out equally between 6 people. How much does each one receive?

18 May makes 17 necklaces out of 2159 beads. Each necklace has the same number of beads.
How many beads are there on each necklace?

19 A teacher shares out 868 sheets of paper equally amongst her class of 28 students.
How many sheets of paper does each student receive?

20 Havering School has 434 students divided equally into 14 classes.
How many students are in each class?

EXERCISE 1.4

Without using a calculator, work out the answers to the following.

1 $-3 + (-4)$ 2 $9 + (-5)$ 3 $7 - (-1)$ 4 $-6 - (-2)$

5 $-7 + 3$ 6 $-9 - (6)$ 7 $-1 - -3$ 8 $-8 + -5$

9 $(2) - +3$ 10 $-5 + -5$ 11 $3 - -4$ 12 $-4 - +5$

13 $1 + -6$ 14 $-6 - -1$ 15 -2×-8 16 $-7 \times +9$

17 5×-6 18 $-16 \div -4$ 19 $-7 \div 7$ 20 $24 \div -8$

21 Arrange these in order of size, smallest first: 0, −9, 2, 7, −4.

22 Arrange these in order of size, smallest first: −24, −28, 15, 25, −10.

23 What number lies midway between −18 and −32?

24 What number lies one-third of the way from −5 to 25?

EXERCISE 1.5

1 List all the prime numbers from 22 to 65 inclusive.
You should find there are ten such prime numbers altogether.

2 Use your result from question **1** to help answer these questions:
   a) How many primes are there between 40 and 60 inclusive?
   b) What is the next prime number above 55?
   c) Find two prime numbers that add together to make 120.
   d) Write 123 as a product of two prime factors.

3 Use the factor tree method to obtain the prime factorisation of:
   a) 75   b) 84   c) 420

4 Use the factor tree method to obtain the prime factorisation of:
   a) 24   b) 81   c) 192
   What do you notice about all three of your answers?

5 When 250 is written as a product of primes, the result is $5^x \times y$ where x and y are positive integers. Find the values of x and y.

EXERCISE 1.6

1 Use the method of inspection to write down the highest common factor of each pair of numbers.
Check your result in each case.
   a) 15 and 18   b) 48 and 60   c) 30 and 75
   d) 16 and 24   e) 44 and 66   f) 34 and 170

2 Write each of the following numbers as a product of prime factors.
Hence find the highest common factor of each pair of numbers.
   a) 35 and 28   b) 42 and 105   c) 24 and 90
   d) 66 and 220   e) 90 and 105   f) 140 and 175

3 Use Euclid's method to find the highest common factor of each pair of numbers.
   a) 10 and 15   b) 40 and 48   c) 45 and 60
   d) 28 and 42   e) 33 and 55   f) 38 and 190

EXERCISE 1.7

Find the lowest common multiple (LCM) of each of these pairs of numbers.
You may use whatever method you prefer.

1 15 and 25   **2** 18 and 27   **3** 10 and 35   **4** 14 and 26

5 4 and 34   **6** 21 and 28   **7** 22 and 33   **8** 18 and 24

9 30 and 45   **10** 36 and 45   **11** 42 and 63   **12** 20 and 50

13 75 and 175   **14** 69 and 92   **15** 39 and 65   **16** 8 and 32

17 a) Write 90 and 210 as products of their prime factors.
    b) Hence find the LCM of 90 and 210.

18 a) Write 93 and 124 as products of their prime factors.
    b) Hence find the LCM of 93 and 124.
    c) Find also the HCF of 93 and 124.

19 a) Write 6, 9 and 15 as products of their prime factors.
    b) Work out the HCF of 6, 9 and 15.
    c) Work out the LCM of 6, 9 and 15.

20 a) Write 25, 30 and 80 as products of their prime factors.
    b) Work out the HCF of 25, 30 and 80.
    c) Work out the LCM of 25, 30 and 80.

21 Michael has two friends who sometimes come round to his house. Keith comes round once
every 3 days and Peter comes round once every 7 days.
How often are both friends at Michael's house together?

22 Christine owns three sports cars. She cleans the Ferrari once every 6 days, the Lotus once every
14 days and the Porsche once every 21 days. Today she has cleaned all three cars.
When will she next clean all three cars on the same day?

CHAPTER 2

Fractions and decimals

 EXERCISE 2.1

Write these fractions in their simplest terms.

1 $\frac{12}{48}$ **2** $\frac{24}{72}$ **3** $\frac{8}{20}$ **4** $\frac{22}{77}$ **5** $\frac{5}{35}$ **6** $\frac{42}{84}$

7 $\frac{50}{175}$ **8** $\frac{39}{45}$ **9** $\frac{72}{96}$ **10** $\frac{18}{27}$ **11** $\frac{10}{35}$ **12** $\frac{42}{70}$

Arrange these fractions in order of size, smallest first.

13 $\frac{3}{4}, \frac{5}{6}, \frac{7}{10}, \frac{11}{15}$ **14** $\frac{4}{9}, \frac{1}{3}, \frac{1}{4}, \frac{5}{12}$ **15** $\frac{2}{5}, \frac{1}{6}, \frac{3}{10}, \frac{4}{15}$

Work out the answers to these without using a calculator.

16 $\frac{1}{5} + \frac{1}{2}$ **17** $\frac{3}{8} - \frac{1}{3}$ **18** $\frac{1}{4} + \frac{3}{7}$ **19** $\frac{5}{6} - \frac{2}{3}$

20 $\frac{1}{6} + \frac{2}{5}$ **21** $\frac{3}{5} + \frac{1}{4}$ **22** $\frac{5}{7} - \frac{1}{3}$ **23** $\frac{2}{5} + \frac{2}{3}$

24 $\frac{3}{7} - \frac{2}{5}$ **25** $\frac{5}{8} - \frac{1}{3}$

 Use pencil and paper to work out the answer to each of these. Then use your calculator fraction key to check your answers.

 26 $\frac{1}{3} + \frac{1}{4} - \frac{1}{2}$ **27** $\frac{3}{5} - \frac{1}{3} + \frac{1}{4}$ **28** $\frac{2}{3} + \frac{1}{2} - \frac{3}{4}$ **29** $1\frac{2}{3} + 3\frac{5}{7}$ **30** $2\frac{3}{4} + 1\frac{5}{9}$

31 $4\frac{1}{5} - 2\frac{3}{4}$ **32** $5\frac{3}{10} - 3\frac{5}{6}$ **33** $10\frac{7}{9} + 6\frac{5}{6}$ **34** $3\frac{2}{7} - 1\frac{3}{5}$ **35** $5\frac{1}{6} - 2\frac{3}{8}$

 EXERCISE 2.2

Work out these multiplications and divisions. Show all your working clearly.

1 $\frac{3}{4} \times \frac{2}{5}$ **2** $\frac{2}{3} \times \frac{5}{7}$ **3** $\frac{2}{7} \times \frac{5}{8}$ **4** $\frac{4}{9} \times \frac{15}{22}$

5 $\frac{7}{8} \times 24$ **6** $\frac{3}{11} \times \frac{7}{9}$ **7** $\frac{3}{10} \times \frac{5}{6}$ **8** $35 \times \frac{3}{5}$

9 $\frac{3}{7} \div \frac{9}{10}$ **10** $\frac{3}{5} \div \frac{6}{7}$ **11** $\frac{5}{6} \div \frac{8}{9}$ **12** $\frac{5}{8} \div 40$

13 $\frac{2}{3} \div \frac{8}{11}$ **14** $36 \div \frac{4}{5}$ **15** $\frac{1}{5} \div \frac{1}{20}$ **16** $\frac{8}{9} \div \frac{18}{19}$

17 $\frac{6}{35} \times \frac{14}{15}$ **18** $\frac{3}{11} \div \frac{3}{11}$ **19** $\frac{5}{7} \times \frac{14}{15}$ **20** $\frac{1}{5} \times \frac{5}{6} \div \frac{1}{3}$

Work out the answers to these calculations. Remember to convert mixed fractions into top-heavy fractions before doing the multiplications or divisions.

21 $5\frac{3}{7} \times 1\frac{9}{19}$ **22** $4\frac{1}{6} \times \frac{4}{15}$ **23** $1\frac{5}{6} \times 1\frac{3}{11}$ **24** $2\frac{2}{7} \times 1\frac{11}{24}$

25 $1\frac{3}{4} \div 2\frac{1}{7}$ **26** $\frac{14}{15} \div 1\frac{2}{5}$ **27** $3\frac{1}{5} \div 2\frac{2}{3}$ **28** $5 \div 2\frac{2}{3}$

29 $2\frac{1}{2} \div 1\frac{2}{3}$ **30** $2\frac{3}{4} \times 5\frac{1}{3}$

Use pencil and paper to work out the answer to each of these. Then use your calculator fraction key to check your answers.

31 $\frac{12}{25} \times 1\frac{2}{3}$ **32** $6\frac{2}{3} \times 1\frac{1}{5}$ **33** $3\frac{3}{8} \div 2\frac{1}{4}$ **34** $25 \div 1\frac{3}{7}$

35 $33\frac{1}{3} \times \frac{9}{20}$ **36** $2\frac{6}{11} \div 2\frac{7}{22}$

 EXERCISE 2.3

Work out the answers to these calculations without a calculator.

1 6.27×0.3 **2** 5.09×0.2 **3** 2.8×3.6 **4** 5.7×1.4

5 7.14×3.8 **6** 6.71×9.2 **7** 0.04×0.1 **8** 0.7×0.02

9 $44.45 \div 7$ **10** $2.675 \div 0.5$ **11** $30.66 \div 2.1$ **12** $0.36 \div 0.9$

13 $29.12 \div 1.3$ **14** $0.6413 \div 0.01$ **15** $30.566 \div 3.1$ **16** $20.91 \div 0.17$

17 Given that $28 \times 64 = 1792$, work out the values of
   a) 6.4×2.8 **b)** 0.64×28 **c)** 640×280

18 Given that $215 \times 38 = 8170$, work out the values of
   a) 2.15×3.8 **b)** 21.5×0.38 **c)** $8.17 \div 0.38$

19 Given that $518 \times 74 = 38\,332$, work out the values of
   a) 51.8×7.4 **b)** 0.518×7.4 **c)** $383.32 \div 7.4$

20 Use the fact that $306 \div 17 = 18$ to work out:
   a) 18×1.7 **b)** $306 \div 0.17$ **c)** 1.8×0.17

 EXERCISE 2.4

Write these terminating decimals as exact fractions. Give each answer in its lowest terms.

1 0.36 **2** 0.05 **3** 0.7 **4** 0.76 **5** 0.0375

6 0.55 **7** 2.47 **8** 3.33 **9** 0.65 **10** 1.102

Write these recurring decimals as exact fractions, in their lowest terms. Show your method clearly.

11 $0.\dot{6}$ **12** $0.3\dot{7}$ **13** $5.\dot{1}$ **14** $0.\dot{4}5\dot{2}$

15 $0.13\dot{4}$ **16** $7.\dot{3}$ **17** $0.2\dot{4}$ **18** $6.3\dot{9}\dot{6}$

Write these fractions as decimals.

19 $\frac{3}{5}$ **20** $\frac{7}{11}$ **21** $\frac{5}{9}$ **22** $\frac{13}{30}$

23 Which is larger: $5.06\dot{9}$ or 5.1?

24 Aaron says, '$2.\dot{5}$ is exactly twice as big as $1.2\dot{5}$.'
   Is Aaron right or is he wrong? Explain your answer carefully.

 ## EXERCISE 2.5

Round the following decimal numbers to the indicated number of decimal places.

1 5.518 49 (3 d.p.) **2** 26.037 2 (2 d.p.) **3** 9.751 (1 d.p.) **4** 1.289 74 (4 d.p.)

5 83.666 (2 d.p.) **6** 52.40 (1 d.p.) **7** 9.340 071 1 (4 d.p.) **8** 0.3498 (3 d.p.)

Round the following to the specified number of significant figures.

9 67.07 (3 s.f.) **10** 25.32 (1 s.f.) **11** 183 (2 s.f.) **12** 78.984 91 (4 s.f.)

13 9134 (2 s.f.) **14** 18.4 (1 s.f.) **15** 0.056 17 (3 s.f.) **16** 129.51 (2 s.f.)

17 The lengths of pieces of wood are measured as 27 cm, to the nearest cm.
   Five pieces of wood are placed in a straight line. Work out the smallest possible **total** length.

18 A rectangle has length 57 mm and width 8 mm, both correct to the nearest mm. Work out the
   largest possible area it could have, giving your answer correct to 3 significant figures.

19 Last week, the number attending the local rugby match was 6400, correct to the nearest
   hundred. The week before it was 6350, correct to the nearest ten. Is it possible that the same
   number of people attended in both weeks? Explain your answer.

20 Jerry says that 48 998 when rounded to the nearest thousand is twice as much as 24 499 when
   rounded to the nearest thousand. Is Jerry correct? Explain why.

CHAPTER 3

Ratios and percentages

EXERCISE 3.1

Express each of these ratios in its simplest form.

1 6 : 9   **2** 18 : 12   **3** 25 : 10   **4** 48 : 40   **5** 20 : 35 : 40

6 60 : 100 : 140   **7** 14 : 49 : 63   **8** 55 : 88 : 220   **9** 45 : 60 : 75   **10** 18 : 45 : 81

11 Share £150 in the ratio 3 : 5 : 7.

12 Share 120 in the ratio 7 : 9 : 14.

13 Share 60 cm in the ratio 2 : 3 : 7.

14 Share $70 in the ratio 2 : 5 : 7.

15 Share 90 kg in the ratio 1 : 6 : 8.

16 Share £520 in the ratio 3 : 4 : 6.

17 Pete and Lian share some nuts.
Pete takes 5 nuts for every 3 nuts that Lian takes.
There are 48 nuts.
How many **more** nuts does Pete take than Lian?

18 Jean has three daughters.
Michelle is 20 years old, Deborah is 22 years old and Lisa is 24 years old.
Write the ratio of their ages. Give your answer in its simplest form.

19 Each month Marc spends his money on food, rent and other expenses in the ratio 2 : 7 : 8
In January he spent £350 on rent.
a) Work out how much he spent on food in January.
b) Work out how much he spent in total in January.

20 Which of these is the best buy?

15 exercise books price £3.00

24 exercise books price £5.40

21 Here are the ingredients for making a vegetable soup for 12 people:

    6 carrots
    4 onions
    2400 ml stock
    150 g lentils
    12 g thyme

Work out how much of each ingredient is needed to make vegetable soup for 18 people.

EXERCISE 3.2

Write these percentages as fractions in their lowest terms.

1 65%   **2** 90%   **3** 36%   **4** $7\frac{1}{2}$%

5 12%   **6** 340%   **7** 12.5%   **8** $66\frac{2}{3}$%

Express these fractions as percentages. You should do them by hand first.
Then use your calculator's fraction key to check each one.

9 $\frac{28}{50}$ **10** $\frac{15}{20}$ **11** $\frac{57}{60}$ **12** $\frac{12}{30}$ **13** $\frac{24}{60}$ **14** $\frac{22}{55}$ **15** $\frac{12}{40}$ **16** $\frac{36}{80}$

17 Arrange these quantities in order of size, smallest first.

     0.42, $\frac{4}{11}$,   45%, $\frac{3}{7}$,   0.4,   41%

18 Harry buys a book for £40 and sells it for £32.
   a) How much money has he lost?
   b) What is his percentage loss?

19

Steve's Sausages	**Bertie's Bangers**
72% beef	44 g out of every 60 g beef

Rob wants to buy sausages that contain the most beef.
Which sausages should he buy? Explain why.

20 Maria buys a house for £240 000 and sells it three years later for £300 000.
What is her percentage profit?

EXERCISE 3.3

1 Increase 360 by 18%.   **2** Decrease 72 400 by 15%.

3 Increase 1500 by 20%, then by a further 87% of its new value.

4 Jon buys a DVD recorder. How much VAT does he pay?

> Special Offer!
> DVD recorder
> only £290
> plus 17.5% VAT

5 In his first Mathematics test Leo scored 60 marks. In the second test he scored 70 marks.
Calculate the percentage increase in his test marks.

6 Lucy bought a new car. She paid £22 000. Each year the value of the car depreciated by 12% of its value at the beginning of that year. Work out the value of Lucy's car:
   a) after one year
   b) after four years.

7 44% of the students in a school are under 14 years of age.
If there are 650 students in the school, work out how many students in the school are 14 years old or more.

8 Sarah's pocket money this year is £40 per month. Next year she is getting a 22% rise, but 15% of her new pocket money has to be put into a savings account.
   a) Work out Sarah's monthly pocket money for next year.
   b) Work out how much she has to put into a savings account each month next year.

9 Pete bought some items on the internet and then sold them at a garage sale.
What percentage profit or loss has he made on each of these items?
   a) Boots bought for £60, sold for £45
   b) A CD collection bought for £80, sold for £90

10 John bought a necklace for £120. He sells the necklace to Mia at a profit of 35%.
Mia then sells the necklace to Lily at 45% more than what she paid John for the necklace.
Calculate how much Lily paid for the necklace.

EXERCISE 3.4

This exercise contains a mixture of forwards and reverse percentage problems – so you can train yourself to spot the difference! You should use multiplying factor methods as much as possible.

1 A book shop reduces the prices of all its books by 15% in a sale.
    a) Find the sale price of a book originally priced at £28.
    b) Find the original price of a book that is £17 in the sale.

2 Mr Jones is left with a monthly income of £2028 after 35% deductions have been made. Work out Mr Jones's monthly salary before any deductions.

3 A pair of sunglasses costs £105.60 inclusive of VAT at 17.5%.
Jerry is told that he does not have to pay the VAT.
Work out how much Jerry should pay.

4 A diet organisation says that you will lose at least 16% of your body weight in 3 months on their special Lighterdiet.
    a) Pat weighs 90 kg. What is the least she should expect to lose if she goes on the Lighterdiet for 3 months?
    b) Tilly weighs 67.2 kg after losing 16% of her body weight on the Lighterdiet.
        Work out how much she weighed at the start of her diet.

5 A football stadium increases its seating capacity by 20%. It can now seat 1800 people.
Work out its seating capacity before the increase.

6 Sanjay bought a house. He sold it 4 years later for £85 000 after making a profit of 25%.
Work out how much Sanjay paid for the house.

7 The number of students in a school increases by 10%.
As a result the number of teachers the school employs increases by $12\frac{1}{2}$%.
The school had 120 teachers before the increase and has 1210 students after the increase.
    a) Work out how many teachers will be employed after the increase.
    b) Work out how many students were in the school before the increase.

8 A playground has a rectangular sandpit. The council has the sandpit made into a larger rectangle. The width of the rectangle is to be increased by 15% to 2.76 m. The length of the sandpit was 4 m but is to be increased by 25%.
    a) Work out the dimensions of the sandpit before it was made larger.
    b) Work out the area of the sandpit after it was made larger.

9 A year ago I bought a new car. It depreciates at 24% per year.
My car is presently worth £3800.
    a) Work out the value of my car when I bought it.
    b) Work out how much my car will be worth three years from now.
        Give your answer to the nearest pound (£).

10 The sides of a triangle are increased by 40%.
  a) Work out the value of h.
  b) Work out the value of b.
  c) By what percentage has the area of the triangle been increased?

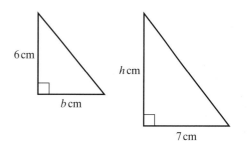

EXERCISE 3.5

1 Work out the simple interest on £2400 invested at 4% per annum for 6 years.

2 Work out the compound interest on £360 invested at 5% per annum for 7 years.

3 Work out the simple interest on £80 invested at 12.5% per annum for 5 years.

4 Work out the compound interest on £8000 invested at 6% per annum for 3 years.

5 £500 is invested at x% compound interest.
  After 5 years it has grown to £701 to the nearest pound.
  Use a trial and improvement method to find the value of x.

6 William invests some money at a compound interest rate of 10% per annum.
  After 4 years his investment is worth £585.64.
  How much did William invest?

7 Prakesh receives £2100 for his 21st birthday. His father says he will look after the money and add 3.5% compound interest to the amount each year.
  How old will Prakesh be when this amount of money has increased to £2862?

8 A ten-year savings bond pays 2% interest for the first year, then 4.5% per annum compound interest after that.
  Work out the total final value after ten years of an initial investment of £1000.

9 Wendy is investing £300 in a long-term savings scheme.
  She can choose between two savings plans.
  Savings Plan A: **Simple** interest at 6% per annum.
  Savings Plan B: **Compound** interest at 5% per annum.
  a) Which plan gives the better return over 3 years?
  b) Which plan gives the better return over 10 years?
  Justify your answers with numerical calculations.

10 Jasmine invests £100 in a savings account at 3% compound interest.
  How many years will it take for Jasmine's investment to exceed £180?

CHAPTER 4

Powers, roots and reciprocals

EXERCISE 4.1

Write down the values of the following, without using a calculator.

1 3^2	**2** 5^3	**3** 6^2	**4** 1^3
5 8^2	**6** 6^3	**7** 10^2	**8** 0^3
9 $\sqrt{25}$	**10** $\sqrt{100}$	**11** $\sqrt[3]{8}$	**12** $\sqrt{4}$
13 $\sqrt{121}$	**14** $\sqrt[3]{27}$	**15** $\sqrt{49}$	**16** $\sqrt[3]{1}$

Use your calculator to work out the values of the following expressions.
Round your answers to 3 significant figures where appropriate.

17 21^2	**18** 3.7^2	**19** 12.9^2	**20** 8^3
21 11.4^3	**22** 2.6^3	**23** $\sqrt{20}$	**24** $\sqrt[3]{500}$
25 $\sqrt[3]{26.1}$	**26** $\sqrt{47.3}$	**27** $\sqrt{7.18}$	**28** $\sqrt[3]{1.9}$

29 Find x if $x^2 = 250$. Give your answer to 3 s.f.

30 Find y if $y^3 = 250$. Give your answer to 2 s.f.

EXERCISE 4.2

Without using a calculator, find the exact values of these.

1 2^6	**2** $27^{\frac{1}{3}}$	**3** 5^3	**4** $100^{\frac{1}{2}}$
5 2^3	**6** 4^4	**7** $8000^{\frac{1}{3}}$	**8** $900^{\frac{1}{2}}$
9 3^5	**10** $0.0001^{\frac{1}{4}}$	**11** 6^3	**12** $32^{\frac{1}{5}}$

Use your calculator to find the value of each expression. Round your answers as indicated.

13 8^5 (3 s.f.)	**14** 7.2^3 (4 s.f.)	**15** 1.4^6 (2 d.p.)	**16** 0.53^4 (4 d.p.)
17 $18^{\frac{1}{4}}$ (3 s.f.)	**18** $51^{\frac{1}{2}}$ (3 s.f.)	**19** $1752^{\frac{1}{5}}$ (3 d.p.)	**20** $0.923^{\frac{1}{3}}$ (3 s.f.)

EXERCISE 4.3

Without using a calculator, write the following expressions as simply as possible.

1 $9^{\frac{3}{2}}$

2 $125^{\frac{2}{3}}$

3 $32^{\frac{3}{5}}$

4 $1000^{\frac{4}{3}}$

5 $1^{\frac{8}{3}}$

6 $64^{\frac{1}{2}}$

7 $16^{\frac{7}{4}}$

8 $0.008^{\frac{2}{3}}$

9 $27^{\frac{4}{3}}$

10 $4^{\frac{5}{2}}$

11 $100^{\frac{3}{2}}$

12 $216^{\frac{2}{3}}$

EXERCISE 4.4

Work out the values of these, leaving your answers as exact fractions.

1 2^{-3}

2 7^{-2}

3 1^{-6}

4 3^{-4}

5 5^{-3}

6 10^{-2}

7 8^{-1}

8 4^{-3}

9 9^{-1}

10 30^{-2}

Evaluate these expressions, giving your answers as exact fractions.

11 $\left(\frac{8}{7}\right)^{-1}$

12 $\left(\frac{10}{17}\right)^{-1}$

13 $\left(\frac{3}{2}\right)^{-3}$

14 $\left(\frac{5}{6}\right)^{-2}$

15 $\left(\frac{1}{3}\right)^{-4}$

16 $\left(\frac{81}{49}\right)^{-\frac{1}{2}}$

17 $\left(\frac{4}{11}\right)^{-1}$

18 $\left(\frac{1000}{27}\right)^{-\frac{1}{3}}$

19 $\left(\frac{64}{125}\right)^{-\frac{2}{3}}$

20 $\left(\frac{16}{9}\right)^{-\frac{3}{2}}$

EXERCISE 4.5

Simplify each of these expressions, giving your answer as a number to a single power.

1 $7^5 \times 7^9$

2 $2^{10} \times 2^6$

3 $10^8 \div 10^2$

4 $5^{18} \div 5^6$

5 $8^0 \times 8^5$

6 $9^9 \times 9$

7 $15^{80} \div 15^0$

8 $37^{14} \div 37^7$

9 $(2.6^3)^5$

10 $58^{\frac{7}{2}} \times 58^{\frac{3}{2}}$

11 $\dfrac{6^{15}}{6^3}$

12 $\dfrac{9^{\frac{5}{3}}}{9^{\frac{2}{3}}}$

Work out each of these, giving your answer as an ordinary number.

13 $3^3 \times 3$

14 $7^5 \div 7^3$

15 $9^9 \div 9^8$

16 5×5^2

17 $4^0 \times 4^3$

18 $2^8 \div 2^2$

19 $(2^3)^4$

20 $8^4 \div 8^2$

21 $(6^5)^0$

22 $(3^2)^2$

23 $1^8 \times 1^3$

24 $(10^3)^2$

EXERCISE 4.6

Write these numbers in standard index form.

1 276 000 **2** 1280 **3** 73 000 **4** 9 000 000 **5** 0.072

6 0.004 63 **7** 0.61 **8** 75.2 **9** 287.03 **10** 0.000 01

Write these as ordinary numbers.

11 7.2×10^4 **12** 3.91×10^3 **13** 6×10^{-2} **14** 4.05×10^{-1} **15** 5.73×10^5

16 8.2×10^{-4} **17** 9.987×10^2 **18** 1.04×10^{-3} **19** 2×10^{-5} **20** 6.7673×10^4

EXERCISE 4.7

Work out the answers to these calculations without using a calculator.
Give your answers in standard form.

1 $(2 \times 10^4) + (5 \times 10^3)$ **2** $(3.9 \times 10^2) + (2.7 \times 10^4)$

3 $(8 \times 10^5) - (8 \times 10^4)$ **4** $(2 \times 10^6) - (5 \times 10^7)$

5 $(3.6 \times 10^{12}) \times (2 \times 10^4)$ **6** $(5 \times 10^9) \times (7 \times 10^{-5})$

7 $(42 \times 10^{15}) \div (6 \times 10^7)$ **8** $(4 \times 10^3) \div (8 \times 10^{-6})$

Use your calculator to evaluate these expressions.
Give your answers in standard form, correct to 3 significant figures.

9 The mass of a neutron is 1.675×10^{-24} grams.
Work out the mass of 8.5×10^{14} neutrons.

10 $(7.13 \times 10^{10}) \times (3.28 \times 10^3)$ **11** $\dfrac{(7.4 \times 10^6) + (1.82 \times 10^7)}{(4.56 \times 10^9)}$

12 $0.0472 \div (5.3 \times 10^{-8})$ **13** $(2.6 \times 10^3)^3$

14 $\dfrac{(5.1 \times 10^{-2}) - (3.4 \times 10^{-1})}{(9.76 \times 10^{-7})}$ **15** $(3.6 \times 10^{12}) \div (2 \times 10^4)$

CHAPTER 5

Working with algebra

EXERCISE 5.1

1 If $c = 5$, $h = 7$ and $n = 3$, find the value of:
   a) $2h - 3n$ **b)** $c^2 - 5n$ **c)** $ch - 2hn$ **d)** $(3c - 2h)^2$

2 If $e = 2$, $f = 4$ and $g = -3$, find the value of:
   a) $(e + f)^2$ **b)** $2f - g$ **c)** $3g + 2ef$ **d)** $(3e - g)^2$

3 If $x = -6$, $y = -1$ and $z = 9$, find the value of:
   a) $y^2 + 2z$ **b)** $z(x - 10y)$ **c)** $7(2x + 3z)$ **d)** $\dfrac{x + 2z}{2y}$

4 The final velocity, v, of a particle is found using the formula $v = u + at$
   where u is the initial velocity, t is the time and a is the acceleration.
   a) Use your calculator to work out the value of v when $a = 9.8$, $t = 3.57$ and $u = 2.43$.
   Write your answer correct to 3 significant figures.
   b) Find the value of a when $v = 29$, $u = 5$ and $t = 4$.

5 The distance s, travelled by a particle is given by the formula $s = ut + \frac{1}{2}at^2$
   where u is the initial velocity, t is the time and a is the acceleration.
   a) Use your calculator to work out the value of s when $a = 8.5$, $t = 6.5$ and $u = 1.7$.
   Write your answer correct to 3 significant figures.
   b) Use your calculator to work out the value of u when $s = 29.3$, $t = 4$ and $a = -2$.
   Write your answer correct to 3 significant figures.

6 The time T travelled is given by the formula $T = \dfrac{D}{S}$.
   a) Find the exact value of T when $D = 322$ and $S = 23$.
   b) Find the value of D when $T = 5.2$ and $S = 63.5$.

EXERCISE 5.2

Write these expressions using indices.

1 $x \times x \times x \times x$ **2** $y \times y \times y$ **3** $c \times c$

4 $t \times t \times t \times t \times t$ **5** $3 \times h \times h \times h \times h$ **6** $7 \times t \times t$

Simplify these expressions using the index laws $x^a \times x^b = x^{a+b}$ and $x^a \div x^b = x^{a-b}$.

7 $x^4 \times x^3$ **8** $y^{15} \div y^5$ **9** $z \times z^8 \times z^2$

10 $4x^2 \times 6x^5$ **11** $8x^9 \times 3x$ **12** $40y^6 \div 5y^3$

13 $2y^5 \times 5y^2$ **14** $27z^{27} \div 9z^9$ **15** $3x^4 \times 2x \times 7x^3$

16 $6z^9 \div 12z$

Simplify these expressions using the index law $(x^a)^n = x^{a \times n}$.

17 $(x^4)^2$ **18** $(y^8)^3$ **19** $(3z^5)^3$ **20** $(9x^6)^2$ **21** $(y^{12})^5$

22 $(5z^5)^3$ **23** $(7x^4)^2$ **24** $(2x^9)^3$ **25** $(10x^5y^3)^4$ **26** $(4x^{10}y)^2$

Simplify these expressions.

27 $7x^5 \times 5x^3$ **28** $60y^7 \div 15y^2$ **29** $(8z^5)^2$

30 $35y^2 \div 5y$ **31** $(6x^8)^2$ **32** $32x^9 \div 8x^7$

33 $3x \times 12x^4 \times 2x^2$ **34** $(4x^2y^7)^2$ **35** $(x^{12}y^{11})^4$

36 $3x \times 15x^4$

EXERCISE 5.3

Multiply out the brackets. Simplify the results.

1 $3(x + 7) + 4(x + 1)$ **2** $8(x + 2) + 5(x - 2)$ **3** $9(x + 1) + 5(x + 2)$

4 $6(5x - 2) + 9(2x + 3)$ **5** $8(3x + 2) + 2(4x + 3)$ **6** $5(3x + 2) + 4(3x - 2)$

7 $6(x + 6) + 7(3x + 7)$ **8** $7(x - 1) + 4(2x - 1)$ **9** $4(3x + 2) + 5(2x + 3)$

10 $9(8x - 2) + 3(3x + 4)$

Multiply out the brackets. Simplify the results.
Take special care when there is a negative number in front of the second bracket.

11 $5(6x - 1) + 8(2x - 5)$ **12** $2(6x + 5) - (8x + 9)$ **13** $7(x + 9) + 3(x + 2)$

14 $4(4x + 3) - 2(x - 8)$ **15** $3(5x + 7) - 2(3x + 5)$ **16** $8(x + 2) + 5(x - 7)$

17 $9(3x - 4) - 9(2x - 7)$ **18** $6(x - 1) - 4(5x + 2)$ **19** $10(x + 1) + 7(8x - 1)$

20 $11x - 6(3x - 7) + 5x$

EXERCISE 5.4

Expand and simplify these products of brackets.

1 $(x + 1)(5x + 3)$　　**2** $(6x + 5)(x + 2)$　　**3** $(x + 7)(x + 9)$　　**4** $(3x + 2)(2x - 3)$

5 $(x - 5)(4x + 1)$　　**6** $(5x - 3)(6x - 1)$　　**7** $(9x - 2)(3x + 2)$　　**8** $(2x - 9)(5x - 1)$

9 $(4x + 3)(7x - 2)$　　**10** $(x + 10)(8x - 3)$　　**11** $(7x - 1)(x - 6)$　　**12** $(8x + 5)(2x - 5)$

13 $(6x - 11)(3x - 1)$　　**14** $(x - 8)(x - 4)$　　**15** $(3x + 7)(3x - 7)$　　**16** $(x - 9)(x + 8)$

17 $(x - 5)(x + 5)$　　**18** $(5x - 1)(5x + 1)$　　**19** $(x - 1)^2$　　**20** $(2x + 7)^2$

EXERCISE 5.5

Factorise these expressions. They may all be done using the common factor method.

1 $x^2 - 3x$　　**2** $3x^2 - 15x$　　**3** $6x^2 + 4x$　　**4** $y^2 + 7y$

5 $3y^2 - 15y$　　**6** $8x + 12x^2$　　**7** $20y^2 - 12y$　　**8** $20y^2 - 12$

9 $5fg + g^2$　　**10** $6y^2 - 8y$　　**11** $3x^3 - 2x^2$　　**12** $12x^8 + 14x^7$

13 $8ab - 27b^2$　　**14** $10y - 25$　　**15** $6y + 3xy$　　**16** $21x^2y + 14xy^2$

17 $24y^3 - 16y^2$　　**18** $14y - 7y^3$　　**19** $9x^2 + 3$　　**20** $4p^2q - 6pq^2 + 8pq$

EXERCISE 5.6

Factorise these quadratic expressions.

1 $x^2 + 5x + 4$　　**2** $x^2 + 15x + 14$　　**3** $x^2 + 8x + 15$　　**4** $x^2 + 10x + 21$

5 $x^2 + 7x + 12$　　**6** $x^2 - 2x + 1$　　**7** $x^2 - 9x + 14$　　**8** $x^2 - 12x + 27$

9 $x^2 - 14x + 40$　　**10** $x^2 - 9x + 18$　　**11** $x^2 + 5x - 14$　　**12** $x^2 - x - 20$

13 $x^2 - 5x - 24$　　**14** $x^2 + 3x - 10$　　**15** $x^2 - 11x + 18$　　**16** $x^2 + 10x + 24$

17 $x^2 + 4x - 12$　　**18** $x^2 - 8x - 20$　　**19** $x^2 - 13x + 22$　　**20** $x^2 - 2x - 35$

EXERCISE 5.7　

Factorise these quadratic expressions.

1 $2x^2 + 11x + 5$　　**2** $3x^2 + 5x + 2$　　**3** $2x^2 + 9x + 7$　　**4** $5x^2 + 7x + 2$

5 $7x^2 + 13x - 2$　　**6** $3x^2 - 2x - 5$　　**7** $5x^2 + 4x - 1$　　**8** $7x^2 - 5x - 2$

9 $2x^2 - 7x + 6$ **10** $3x^2 + 8x - 3$ **11** $2x^2 - 9x + 10$ **12** $8x^2 - 14x + 3$

13 $6x^2 + 11x + 4$ **14** $6x^2 - 17x + 5$ **15** $3x^2 + 17x + 10$ **16** $9x^2 - 12x + 4$

17 $8x^2 + 10x - 3$ **18** $10x^2 - 13x - 3$ **19** $12x^2 + 11x + 2$ **20** $7x^2 - 18x - 9$

EXERCISE 5.8A

Factorise these expressions, using the difference of two squares method.

1 $x^2 - 9$ **2** $x^2 - 100$ **3** $y^2 - 25$ **4** $x^2 - 900$ **5** $2y^2 - 2$

6 $5x^2 - 20$ **7** $3x^2 - 300$ **8** $2y^2 - 72$ **9** $7y^2 - 28$ **10** $10x^2 - 490$

EXERCISE 5.8B

This exercise contains a mixture of all the different factorising methods you have learnt so far.

Factorise these expressions.

1 $x^2 - 5x$ **2** $y^2 + 7y + 10$ **3** $x^2 - 64$ **4** $x^2 - 10x + 16$

5 $y^2 + 3y - 28$ **6** $5x^2 - 15x$ **7** $2x^2 - 11x + 14$ **8** $y^2 - 81$

9 $y^2 - 5y - 50$ **10** $x^2 - 9x - 10$ **11** $x^2 + 16x$ **12** $12x^2 + 4x - 1$

13 $7xy - 21y^2$ **14** $2y^2 - 5y + 2$ **15** $y^2 + y - 30$ **16** $2x^2 - 50$

17 $x^2 - 9x + 18$ **18** $6y^2 + 9y$ **19** $3x^2 - 20x - 7$ **20** $5y^2 - 125$

EXERCISE 5.9

1 A square has sides of length m cm. Obtain a formula for the perimeter, P, of the square.

2 A square has length g cm.
Write down a formula in terms of g for:
a) the perimeter, P, of the square
b) the area, A, of the square.

  g cm

3 A rectangle has width x cm.
Its length is 10 cm more than its width.
Write down a formula in terms of x for:
a) the length, L, of the rectangle
b) the perimeter, P, of the rectangle
c) the area, A, of the rectangle.

 x cm

4 Becky buys n books at £p each.
Obtain an equation for the total cost, C, of the books that Becky buys.

5 Pens cost 5p each and pencils cost 8 pence each.
Nick buys x pens and y pencils.
a) Write down an expression in terms of x and y for the total number of pens and pencils he buys.
b) If the total cost is z pounds, write down an equation for the total cost in terms of x, y and z.

6 The cost of renting a machine is £20 plus a daily charge of £5 per day.
a) Find the cost of hiring the machine for 6 days.
b) Obtain a formula for the cost, £C, of hiring the machine for x days.

7 In a school there are g girls and b boys.
a) Write an expression for the total number of students in the school.
b) Write a formula for the total number of students, N, in the school.

8 On Monday, Melissa runs p kilometres per hour for q hours.
Write down a formula for the total distance, D, that Melissa runs on Monday.

9 The diagram shows a large rectangle divided into four smaller shapes labelled P, Q, R and S.
P is a square with each side c cm and S is a rectangle with base 9 cm and height h cm.
a) Find an expression for the area of rectangle Q.
b) Find a formula for the area, A cm^2, of the large rectangle.

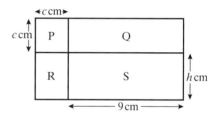

10 A rectangle of dimensions 8 cm by 5 cm has a square of sides x cm cut out from each of its four corners. The sides thus formed are then folded up to make a rectangular tray.
a) Find an expression for the area of the base of the tray.
b) Find a formula for the volume, V cm^3, of the rectangular tray.

EXERCISE 5.10

Rearrange these formulae so that the indicated letter becomes the subject.

1 $A = lw$ (w)

2 $P = 2b + 2h$ (h)

3 $y = 5x - 2$ (x)

4 $D = s \times t$ (t)

5 $E = \dfrac{360}{n}$ (n)

6 $S = 180(n - 2)$ (n)

7 $C = \pi d$ (d)

8 $T = x(y - z)$ (x)

9 $v = u + at$ (t)

10 $m = \dfrac{y}{x}$ (y)

11 $x^2 = 2bh$ (b)

12 $v = u + ft$ (t)

13 $a = \sqrt{x + 3}$ (x)

14 $a^2 = c^2 - b^2$ (c)

15 $y = x^2 + 5$ (x)

16 $y = \dfrac{3}{x}$ (x)

17 $v^2 = u^2 + 2as$ (a)

18 $V = \pi r^2 h$ (r)

19 $A = \frac{1}{2}(a + b)h$ (a)

20 $F = 32 + \dfrac{9C}{5}$ (C)

CHAPTER 6

Algebraic equations

EXERCISE 6.1

Look at the various algebraic statements labelled A to J.

A $7x + 2 = 3$ B $6y + 5$

C πr^2 D $V = L \times B \times H$

E $5x - 2x = 3x$ F $x^3 = 27$

G $2c + 3 = c + 3 + c$ H $A = \frac{1}{2}(a + b)h$

I $x^2 - 9$ J $12x = 30$

1   Which ones are expressions?

2   Which ones are equations?

3   Which ones would you call formulae?

4   Pick out any identities, and rewrite them using the identity sign, $\equiv$.

EXERCISE 6.2

Solve these algebraic equations. Use a formal method, and show the steps of your working.

1   a) $8t = 48$ b) $u - 6 = 4$ c) $\frac{1}{4}p = 5$
    d) $20g = 10$ e) $5q = 1$ f) $w + 8 = 3$

    g) $6y = 5$ h) $y + 12 = 5$ i) $\frac{p}{6} = -3$

2   Find the values of the letters in each of these equations.
    a) $a - 3 = 6$ b) $7b = 77$ c) $5c = 17$
    d) $15 = 28 - d$ e) $12 - e = 12$ f) $8f = 2$
    g) $20 = 30 - g$ h) $h^2 = 36$ i) $4i^2 = 4$

EXERCISE 6.3

Solve these algebraic equations, showing the steps of your working clearly.
All the answers should be integers, but some may be negative.

1   $3x + 2 = x + 12$ 2 $22 + 5x = 6x$ 3 $9x + 4 = 4x + 29$ 4 $x + 5 = 13 - x$

5   $8x + 3 = 48 - x$ 6 $23 - 2x = 33 - 7x$ 7 $5x + 3 = 15 - x$ 8 $7 + 4x = 2x + 9$

9   $8x - 5 = x + 9$ 10 $6x = 42 - x$

Solve these algebraic equations, showing the steps of your working clearly.
Answers should be given as top-heavy fractions or mixed numbers, rather than decimals.

11 $6a + 5 = 3a + 6$ **12** $2b + 7 = 9b - 3$ **13** $8c - 5 = 3c - 2$ **14** $9d + 1 = 17 - 3d$

15 $5e + 12 = e + 3$ **16** $10f - 3 = 11f - 1$ **17** $4g - 9 = 2 + g$ **18** $7h - 2 = 6 + 4h$

19 $i + 4 = 7 - i$ **20** $3j - 4 = 5 + 10j$

EXERCISE 6.4

Multiply out the brackets, and hence solve these equations. Show each step of your working.

1 $6(x - 3) + 2 = 14$ **2** $3(x - 4) = 23 - 2x$

3 $4(x + 10) = 50 - x$ **4** $8(x - 3) = 8$

5 $2(5x + 3) = 7x + 24$ **6** $x + 7 = 5(x - 1)$

7 $4(3x - 5) = 3(2x + 1) + 1$ **8** $5(x - 1) = 3(x - 5)$

9 $7(x + 8) = 4(x + 7) - 11$ **10** $2(5x + 3) = 4(x - 3)$

11 $6(x + 3) - 4(2x - 3) = 10$ **12** $3x - 7 = 6x - 5(2x + 1)$

Write an equation, involving brackets, to formulate each of the problems below.
Then expand your brackets and solve the equation, to obtain the answer to the problem.

13 Ben earned y euros per hour in January.
In February his earnings were increased by 2 euros per hour.
a) Write down an expression, in terms of y, for Ben's new earnings per hour.
b) Write down an expression, using brackets, for Ben's earnings if he works for 10 hours in February.

14 Simon thinks of a number. He multiplies it by 7 and then adds 5 to the new total.
a) Write down an expression for this information. Use n for Simon's mystery number.
Simon gets the same answer when he adds 29 to his mystery number.
b) Write down an equation for this information.
Solve your equation to find Simon's mystery number.

15 Michael and Tanya each think of the same number.
Michael multiplies the number by 4, and then adds 2.
Tanya adds 4 to the number, and then multiplies the answer by 2.
They both end up with the same answer.
a) Write this information as an equation.
b) Solve your equation to find the number they both thought of.

EXERCISE 6.5

Solve these equations.

1 $\dfrac{x+2}{2} = \dfrac{x+1}{3}$

2 $\dfrac{x+2}{5} = 5$

3 $\dfrac{4x+3}{5} = \dfrac{x+3}{2}$

4 $\dfrac{5x}{8} - \dfrac{x}{4} = 1$

5 $\dfrac{x+4}{5} = \dfrac{x}{3}$

6 $\dfrac{7x-4}{3} = x - 3$

7 $\dfrac{x-5}{2} = \dfrac{x-3}{4}$

8 $\dfrac{3x+7}{4} = \dfrac{2x+11}{9}$

9 $\dfrac{x+2}{10} = \dfrac{4-x}{5}$

10 $\dfrac{2x+7}{6} + \dfrac{5x-5}{3} = 0$

EXERCISE 6.6

Use trial and improvement to solve these equations correct to 1 decimal place.

1 The equation $x^2 + 3x = 24$ has a solution between $x = 3$ and $x = 4$.

2 The equation $x^2 + x = 33$ has a solution between $x = 5$ and $x = 6$.

3 The equation $x^3 + x = 4$ has a solution between $x = 1$ and $x = 2$.

4 The equation $x^2 - 5x = 20$ has a solution between $x = 7$ and $x = 8$.

5 The equation $x^3 + 5x = 117$ has a solution between $x = 4$ and $x = 5$.

6 The equation $x^2 - 3x = 50.5$ has a solution between $x = 8$ and $x = 9$.

7 The equation $2x^2 + x = 2$ has a solution between $x = 0$ and $x = 1$.

Use trial and improvement to solve these equations correct to 2 decimal places.

8 The equation $2x^3 - x = 500$ has a solution between $x = 6$ and $x = 7$.

9 The equation $x^2(x + 1) = 34$ has a solution between 2 and 3.

10 The equation $x^2 + 2x = 7$ has a solution between $x = 1$ and $x = 2$.

CHAPTER 7

Graphs of straight lines

EXERCISE 7.1

1 Using the diagram, write down the coordinates of A, B, C, D and E.

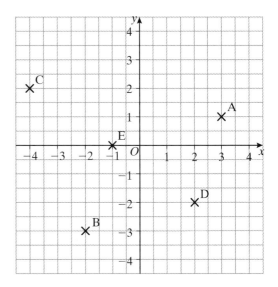

2 The questions refer to the diagram.
   a) Which point is at $(-4, -2)$?
   b) What are the coordinates of L?
   c) Which point is midway between $(2, -6)$ and $(8, 0)$?
   d) What are the coordinates of H?
   e) Which point has the largest y coordinate?
   f) Which point has the smallest x coordinate?
   g) Which point has the same x coordinate and y coordinate?

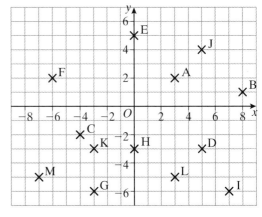

3 Follow these instructions carefully.
   Draw a coordinate grid and label both axes from -10 to 10 on the x axis and -4 to 4 on the y axis.
   Now draw line segments as follows:
   a) From $(-8, 3)$ to $(-8, -3)$ **b)** From $(-2, 3)$ to $(-2, -3)$ **d)** From $(6, 3)$ to $(6, -3)$
   From $(-8, 0)$ to $(-5, 3)$ **c)** From $(0, 3)$ to $(4, 3)$ From $(6, 3)$ to $(9, 3)$
   From $(-8, 0)$ to $(-5, -3)$ From $(2, 3)$ to $(2, -3)$ From $(6, 0)$ to $(8, 0)$
                                                                     From $(6, -3)$ to $(9, -3)$

   You should find that you have named a 4-sided shape.
   Draw a sketch of this shape.

4 Here is a puzzle using a coordinate grid.
Draw a coordinate grid so that x and y both run from -6 to 6.
  a) Draw three squares in the following positions:
  Join $(2, 2)$ to $(-1, 2)$ to $(-1, -1)$ to $(2, -1)$ to $(2, 2)$.
  Join $(-1, 2)$ to $(-4, 2)$ to $(-4, 5)$ to $(-1, 5)$ to $(-1, 2)$.
  Join $(2, -1)$ to $(5, -1)$ to $(5, -4)$ to $(2, -4)$ to $(2, -1)$.
  You should now have three squares.
  b) To solve the puzzle move exactly four of the lines so that there are exactly five squares.

 EXERCISE 7.2

For questions **1** to **8** you are given a linear function and an incomplete table of values.

Work out the missing values to complete the table, and then plot the graph of the corresponding line segment. You may use either graph paper or squared paper.

1 $y = 3x + 1$

x	-2	0	1
y	-5	1	

2 $y = 4x$

x	-1	0	2
y	-4		

3 $y = 3x - 2$

x	-1	0	2
y			

4 $y = 4 - x$

x	-2	0	2
y			

5 $x + y = 5$

x	-5	0	5
y			

6 $y = \frac{1}{2}x + 2$

x			
y			

7 $y = 2x - 1$

x			
y			

8 $2x + 3y = 6$

x			
y			

9 Draw up a set of coordinate axes so that x runs from -3 to 3 and y from -10 to 10.
  a) Calculate the coordinates of three points that lie on the line $y = 3x + 2$.
  Hence plot the line $y = 3x + 2$ on your coordinate axes.
  b) Now calculate the coordinates of three points that lie on the line $y = 3x - 2$.
  Plot the line $y = 3x - 2$ on the same set of coordinate axes.
  c) Look at your two graphs. What do you notice?

10 Draw up a set of coordinate axes so that x runs from -6 to 6 and y from 0 to 12.
  a) Calculate the coordinates of three points that lie on the line $y = 6 - x$.
  Hence plot the line $y = 6 - x$ on your coordinate axes.
  b) Now calculate the coordinates of three points that lie on the line $y = 6 + x$.
  Plot the line $y = 6 + x$ on the same set of coordinate axes.
  c) Look at your two graphs. What do you notice?

EXERCISE 7.3

Find the gradient m and the intercept c for each of the lines marked in questions **1** to **8**.

1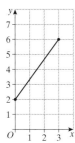

2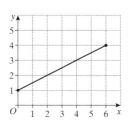

3

4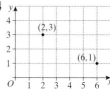

5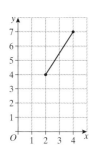

6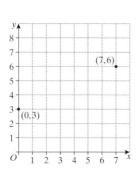

7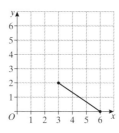

8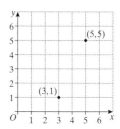

EXERCISE 7.4 pt

1 to 8 Write down the equations of the straight lines whose gradients and intercepts you found in Exercise 7.3, questions **1** to **8**.

9 The diagram shows the graph of a linear function in x.
   a) Find the gradient and intercept of the line.
   b) Hence write down the equation of the straight line.

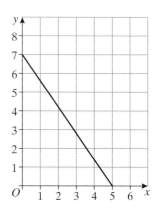

10 The diagram shows the graph corresponding to a linear function of x.

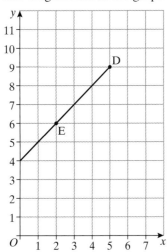

a) Write down the coordinates of the points D and E on the line.
b) Find the gradient and intercept of the line.
c) Hence write down the equation of the straight line.

EXERCISE 7.5

1 Rearrange each of these equations into the form $y = ax + b$.
   Pick out the two that represent a pair of parallel lines.
   a) $x + y = 5$ b) $2x + y = 5$
   c) $y - 8 - x = 0$ d) $x - y + 5 = 0$

2 Look at this list of equations.
   There are four pairs of parallel lines, and one odd one out.

   $y = 3x - 2$ $y = 7x + 3$ $y = 2 + 7x$
   $y = 2x + 3$ $3 - y = 2x$ $y + 5x = 7$
   $y + 2x = 7$ $y - 3x + 4 = 0$ $2y = 3 - 10x$

   a) Pick out the four pairs of parallel lines.
   b) Suggest the equation of another line that is parallel to the odd one out.

3 The line $y = ax + b$ is parallel to $y = 6x - 2$, and passes through the point $(0, -5)$.
   a) Write down the value of a.
   b) Work out the value of b, and hence obtain the equation of the line.

4 The line $y = mx + c$ is parallel to the line $y = 4 - x$, and passes through the point $(2, 0)$.
   a) Find the values of m and c, and write down the equation of the line.
   b) The line also passes through the point $(7, p)$. Find p.

5 A line has equation $\dfrac{y + 3}{4} = x$.

   a) Rearrange the equation into the form $y = mx + c$.
   b) Find the equation of the parallel line that passes through the point $(1, 9)$.

CHAPTER 8

Simultaneous equations

EXERCISE 8.1

Solve these problems using the method of inspection. Write out all the stages clearly.

1 $2x + y = 7$
$x + y = 4$

2 $5x + 2y = 20$
$4x + 2y = 18$

3 $x + 4y = 11$
$x + y = 5$

4 $x + 5y = 9$
$x + 2y = 3$

5 $5x - 2y = 14$
$3x - 2y = 6$

6 $x + 4y = 21$
$x + 2y = 13$

7 $3x + 4y = 42$
$x + 4y = 30$

8 $5x + y = 13$
$5x + 4y = 22$

9 $2x - 3y = 8$
$x - 3y = 1$

10 $2x + y = 9$
$2x + 5y = 13$

EXERCISE 8.2

Solve questions **1** to **8** using the algebraic *addition* method, showing all your working clearly.

1 $4x + y = 9$
$2x - y = 3$

2 $x + y = 8$
$-x + 2y = 13$

3 $3x - y = 1$
$x + y = 3$

4 $3x + y = 9$
$2x - y = 1$

5 $x + y = 7$
$2x - y = 2$

6 $-x + 2y = 7$
$x + 5y = 14$

7 $3x + y = 10$
$2x - y = 10$

8 $x - 2y = 6$
$3x + 2y = 6$

Solve questions **9** to **16** using the algebraic *subtraction* method, showing all your working clearly.

9 $2x + y = 12$
$x + y = 7$

10 $2x + y = 7$
$2x + 3y = 9$

11 $2x + 3y = 8$
$2x + y = -4$

12 $8x + 3y = 23$
$2x + 3y = 17$

13 $5x + y = 14$
$2x + y = 8$

14 $7x + 3y = 23$
$7x + 6y = 32$

15 $3x + 5y = 31$
$x + 5y = 17$

16 $5x + 3y = 26$
$2x + 3y = 14$

Solve questions **17** to **32** using algebra.
For each question you will have to decide whether to add or subtract.

17 $3x + 2y = 8$
$2x - y = 3$

18 $4x + 3y = -1$
$-3x + 2y = 5$

19 $x + 4y = 6$
$3x - 2y = 4$

20 $3x + 5y = 21$
$4x - 2y = 2$

21 $3x + 4y = 23$
$2x + 5y = 20$

22 $4x + 2y = 22$
$x - y = 4$

23 $3x - 2y = -10$
$x + 6y = 0$

24 $2x - 3y = -17$
$3x + y = -9$

25 $5x + 2y = 1$	**26** $2x + 5y = 16$	**27** $5x - 4y = 24$	**28** $4x - 5y = 17$
$8x + 3y = 2$	$x - y = -6$	$2x - y = 9$	$3x - 4y = 13$
29 $2x + 3y = 1$	**30** $5x + 2y = 11$	**31** $8x - 3y = 6$	**32** $3x + 5y = -21$
$3x + 2y = 9$	$3x - 4y = 4$	$2x - y = 1$	$x - 4y = 10$

EXERCISE 8.3

For each of these questions draw a set of coordinate axes on squared paper (or graph paper). Draw the lines corresponding to each equation, and hence solve the simultaneous equations graphically.

1 $x + y = 3$	**2** $y = 2x + 7$	**3** $2x + y = 2$	**4** $y = 2x - 1$
$x + 2y = 5$	$x + y = 1$	$y = x + 5$	$y - x = 2$
5 $3x + 2y = 6$	**6** $y = x + 3$	**7** $y = 2x + 1$	**8** $x + y = 2$
$5x - 2y = 10$	$x + y = 1$	$x - 2y = 4$	$y = -\frac{1}{2}x$

EXERCISE 8.4

Use simultaneous equations to help you solve the following problems.
Remember to show all your working carefully.

1 In a sale, Pete buys 4 pairs of trousers and 3 shirts for £94 and Andy buys 5 pairs of trousers and 6 shirts for £135.50.
  a) Write two simultaneous equations to express this information.
  b) Solve your equations to find the price of a pair of trousers and the price of a shirt.

2 A hire company has a fleet of coaches and minibuses.
2 coaches and 6 minibuses can carry 204 passengers.
3 coaches and 5 minibuses can carry 242 passengers.
  a) Write two simultaneous equations to express this information.
  b) How many passengers can one coach carry?

3 Nick and Diana go to a shop to buy some wool to make a jumper.
Nick buys 3 balls of red wool and 4 balls of white wool for £11.30.
Diana buys 5 balls of red wool and 3 balls of white wool for £12.60.
  a) Write two simultaneous equations to express this information.
  b) Solve your equations to find the cost of a ball of wool in each colour.

4 At a stall, May buys 3 packets of crisps and 5 cans of cola for £4.30, whilst Jerry buys 5 packets of crisps and 4 cans of cola for £5.
  a) Write two simultaneous equations to express this information.
    Explain the meaning of the symbols you use.
  b) Solve your equations to find the cost of a packet of crisps and the cost of a can of cola.

5 On Tuesday some teachers took some students to the local theatre.
Miss Tobias paid £62 for 3 teachers and 20 students.
Mr Angus paid £95 for 5 teachers and 30 students.
Work out the cost of an adult ticket and the cost of a student ticket.

CHAPTER 9

Inequalities

EXERCISE 9.1

Find the whole number (integer) solutions to each of these inequalities.

1 $x > 5$ **2** $2x > 7$ **3** $y < 3$ **4** $4y < 10$

5 $2 < y \leq 7$ **6** $6 \leq 3g \leq 16$ **7** $5 < x < 9$ **8** $8 \leq 2m \leq 23$

9 $3 < f - 2 < 9$ **10** $4 < 3d < 12$ **11** $7 \leq y \leq 10$ **12** $9 \leq 2x \leq 15$

13 $5x \leq 30$ **14** $4p > 13$ **15** $3 < 2x - 1 \leq 17$ **16** $1 \leq w \leq 7$

17 $5 \leq g - 3 \leq 12$ **18** $8 < 5c \leq 20$ **19** $17 \leq p + 2 \leq 18$ **20** $-4 < v < 1$

EXERCISE 9.2

Solve, algebraically, the inequalities in questions **1** to **20**.

1 $x + 8 \geq 20$ **2** $2x + 1 > 15$ **3** $8x - 27 < 13$ **4** $3 - x < 15 + x$

5 $4x - 1 < x + 8$ **6** $5x + 12 \leq 18 + x$ **7** $8x \geq x - 35$ **8** $53 + x < 17 + 3x$

9 $21 - x \leq 11 + 4x$ **10** $4(x + 1) > x + 10$ **11** $x + 19 \geq 29 - 5x$ **12** $7x - 2 < 30$

13 $34 + x > 6 - 5x$ **14** $x - 12 \leq 11 - x$ **15** $8x + 3 \leq x + 3$ **16** $10 - x < 3x + 5$

17 $500 \geq 230 - 9x$ **18** $8 + 5x < 8 - x$ **19** $3(x - 5) \geq x - 8$ **20** $7(3x + 2) < 5(x - 1)$

21 Solve the inequality $\dfrac{x}{8} - 2 \leq 1$ **22** Solve the inequality $\dfrac{2x + 3}{5} > 4$

23 Solve the inequality $\dfrac{4x - 1}{2} > 7$ **24** Solve the inequality $\dfrac{7x + 8}{3} > 2$

EXERCISE 9.3

In questions **1** to **10** you are given the solution to an inequality.
Draw a suitable number line diagram to illustrate the solution in each case.

1 $x < -4$ **2** $x \geq -3$ **3** $2 \leq x \leq 7$ **4** $-4 < x < -1$

5 $x \leq 2$ **6** $x < 10$ **7** $-4 < x < 0$ **8** $-9 < x \leq -4$

9 $-1 \leq x \leq 4$ **10** $x > 3.5$

In questions **11** to **20**, solve each inequality and then illustrate it with a line diagram.

11 $4 + x < 17$ **12** $5x - 1 \leqslant 24$ **13** $18 - x > 8$ **14** $6x + 11 \geqslant 41$

15 $7 < 2x < 24$ **16** $3 - x < 4x - 7$ **17** $5x - 2 \leqslant x + 3$ **18** $7x + 2 < 18 - x$

19 $5 + 2x < x - 1$ **20** $x - 3 \leqslant 3(x + 4)$

EXERCISE 9.4 pt

For each of questions **1** to **5**, draw a coordinate grid in which x and y can range from -5 to 5.

1 Draw the graphs of these straight lines: $x = 1, x = 4, y = 3$ and $y = x$.
Hence shade the region R corresponding to the inequalities: $x \geqslant 1, x < 4, y < 3, y \leqslant x$.

2 Draw the graphs of these straight lines: $x = 4, y = 1, y = 5$ and $y = x - 1$.
Hence shade the region R corresponding to the inequalities: $x < 4, y > 1, y < 5, y > x - 1$.

3 Draw the graphs of these straight lines: $x = -3, y = 2$ and $y = x$.
Hence shade the region R corresponding to the inequalities: $x > -3, y < 2, y > x$.

4 Shade the region R corresponding to the inequalities: $x \geqslant -1, x \leqslant 3, y \geqslant -1, x + y \leqslant 4$.

5 Shade the region R corresponding to the inequalities: $x < 1, y < 0, y < x + 1, x + y > -3$.

6 The diagram shows a region T bounded by three straight lines, P, Q and R.
  a) Write down the equations of the three straight lines, P, Q and R.
     Show clearly which equation applies to which line.
  b) Write down three inequalities that define the region **T**.

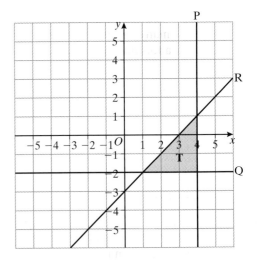

CHAPTER 10

Number sequences

EXERCISE 10.1

In questions **1** to **8**, write down the next two terms in each of the number sequences, and explain how each term is worked out. Give an expression for the nth term in each case.

They are all related to the list of common sequences in the table on page 156 of your textbook.

1 5, 8, 11, 14, 17...

2 4, 9, 14, 19, 24...

3 13, 23, 33, 43, 53...

4 5, 12, 19, 26, 33...

5 0, 3, 8, 15, 24, 35...

6 11, 101, 1001, 10 001, 100 001...

7 2, 8, 18, 32, 50, 72...

8 3, 10, 29, 66, 127, 218...

9 Look at this pattern of trapeziums made from sticks.

| Pattern 1 | Pattern 2 | Pattern 3 |

a) How many sticks would there be in pattern 7?
b) Find a formula for the number of sticks, S, in pattern n.

10 Look at this pattern of spots in rectangles.

   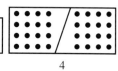

| 1 | 2 | 3 | 4 |

Number of spots in rectangle:

$1 \times 2 = 2$ $2 \times 4 = 8$ $3 \times 6 = 18$ $4 \times 8 = 32$

Number of spots in triangle:

$\dfrac{1 \times 2}{2} = 1$ $\dfrac{2 \times 4}{2} = 4$ $\dfrac{3 \times 6}{2} = 9$ $\dfrac{4 \times 8}{2} = 16$

a) Find an expression for the number of spots in rectangle n.
b) Find an expression for the number of spots in triangle n.

EXERCISE 10.2

1 A number sequence is defined as follows:
- The first term is 28.
- Each new term is 3 less than the previous one.
Use this rule to generate the first five terms of the number sequence.

2 A number sequence is defined as follows:
- The first term is 6.
- To find each new term, double the previous term, and then subtract 5.
Use this rule to generate the first four terms of the number sequence.

3 The nth term of a number sequence is given by the expression $4n - 3$.
a) Write down the first five terms of the number sequence.
b) Work out the 20th term.

4 The nth term of a number sequence is given by the expression $\frac{1}{2}(10n + 3)$.
a) Write down the first six terms of the number sequence.
b) Work out the 10th term.

5 Andrew writes this sequence of numbers: 11, 15, 19, 23, 27...
a) Describe Andrew's pattern in words.
b) Find the tenth term in Andrew's number sequence.

6 The nth term of a number sequence is given by the expression $60 - 2n$.
a) Write down the first five terms of the number sequence.
b) Work out the 50th term.

7 In a certain number sequence, the first term is 5.
Each new term is found by multiplying the previous term by 5.
a) Write down the first five terms of the number sequence.
b) What name is given to this particular number sequence?

8 The nth term of a number sequence is given by the formula $6n + 1$.
a) Work out the first four terms.
b) Find the 20th term.
c) One of the terms in the sequence is 727. Which term is this?

9 The nth term of a number sequence is given by the expression $\dfrac{(n + 1)(n + 2)}{2}$.
a) Write down the first four terms.
b) Work out the 39th term.
c) Explain why all the terms in this sequence are integers.

10 Dana is working with a number sequence.
The nth term of this sequence is given by the expression $8n - 1$.
Dana gets the number 1846 as one of her terms.
Show that she must have made a mistake.

EXERCISE 10.3

1 The first five terms in an arithmetic sequence are:

   8, 14, 20, 26, 32…

   a) Find the tenth term.
   b) Write down, in terms of n, an expression for the nth term of this sequence.

2 The first four terms in an arithmetic sequence are:

   43, 38, 33, 28…

   a) Find the first negative term.
   b) Write down, in terms of n, an expression for the nth term of this sequence.

Here are some arithmetic sequences.
For each one, find, in terms of n, an expression for the nth term of the sequence.

3 5, 12, 19, 26, 33…

4 20, 16, 12, 8, 4…

5 11, 22, 33, 44, 55…

6 23, 20, 17, 14, 11…

7 51, 41, 31, 21, 11…

8 3, 7, 11, 15, 19…

9 Robert has been making patterns with sticks.
   Here are his first three patterns.

     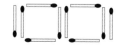 

   Pattern 1   Pattern 2   Pattern 3

   a) Work out the number of sticks in pattern 6.
   b) Give a formula for the number of sticks, s, in pattern n.
   c) Explain how your formula relates to the way the sticks fit together.

10 The seventh term of an arithmetic sequence is 34 and the eighth term is 40.
   a) Write down the common difference for this sequence.
   b) Work out the first term.
   c) Find a formula for the nth term of the sequence.
           Check that your formula works when $n = 7$ and $n = 8$.

CHAPTER 11

Travel and other graphs

1 Uzma cycles to her friend's house.
Here is a description of her journey:
  • Uzma leaves her house at 8 am.
  • She cycles 10 km to arrive at her friend's house at 9.30 am.
  • Uzma spends 75 minutes with her friend.
  • Uzma then cycles back towards her home but stops after 4 km, at 11.15 am to have a rest.
  • She rests for $\frac{1}{4}$ hour.
  • She then cycles straight home, arriving at 12.30 pm.
  a) Draw a travel graph of Uzma's journey.
  b) What is Uzma's average speed, in km per hour, on her way to her friend's house?

2 The diagram shows a vase.
It is in the shape of a
cuboid. Water is poured
into the vase at a steady
rate.

 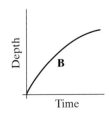 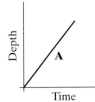 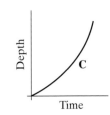

Say which of graphs A, B or C best describes how the depth of water in the vase varies over time. Explain your reasoning.

3 Zac climbs a steep hill.
He gains height at a rate of 5 metres per minute.
It takes him 1 hour to reach the top.
He then stops for 45 minutes to have lunch.
Then he descends, at 10 metres per minute.
  a) How high is the hill?
  b) How long does Zac's descent take?
  c) Copy and complete the graph,
     numbering the scales on both axes.

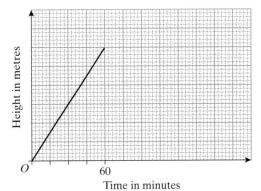

4 Water is poured into a container at a constant rate.
The container is in the shape of a cone, as shown in
the diagram.
Sketch a set of depth/time axes, and complete the
diagram to show how the depth of water in the
container changes over time.

 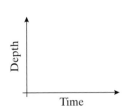

5 Jean saves some money each month.
She saves £20 in the first month and then
saves £30 per month after that.

Number of months (n)	1	2	3	4
Total amount saved (T)	20	50		

  a) Copy and complete the table, to show the
    total amount saved over the first four months.
  b) Find a formula for T in terms of n.
  c) Draw a graph to show how much Jean saves in 1 year.

  Jean needs to save £250.
  d) Use your graph to find out how long it will take Jean to save £250.
  e) Use your formula to work out how many months it will take Jean to save £980.

6 Water runs out of a hole in the bottom of a
container.
The water runs out at a steady rate.
The graph shows how the depth of water in the
container varies over time.

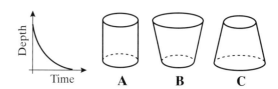

Say which of containers A, B or C best matches
this graph.
Explain your reasoning.

EXERCISE 11.2 pt

1 One Saturday, Maya drives from
Liverpool to Manchester to visit her
grandmother.
She stops along the way to get some
flowers.
On her way home she stops for petrol
and a coffee before returning to
Liverpool.
The travel graph shows her journey.

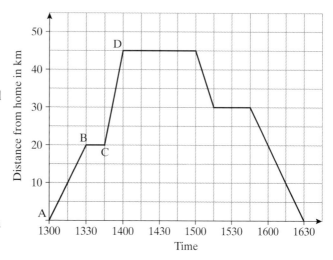

  a) How long does Maya stop for
    flowers?
  b) What time does Maya get to her
    grandmother?
  c) How far is her grandmother from
    Maya's house?
  d) How long did Maya stop for
    petrol and coffee?
  e) Is Maya's speed greater between A and B or between C and D?
    Explain how you can tell.
  f) What time does Maya arrive home?

2 The diagram shows a distance–time graph for a train travelling between Farham and Gradel.

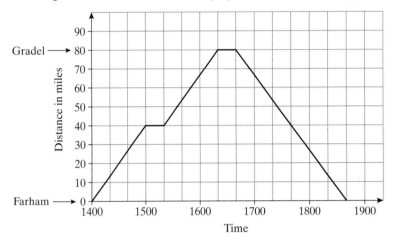

The train leaves Farham at 1400 for its outward journey to Gradel.
The train leaves Gradel at 1640 for its return journey to Farham.
a) Work out the speed of the train on its journey from Gradel to Farham.
b) State one difference between the outward journey and the return journey.
c) State one thing that is the same on the outward journey and the return journey.

At 1500 a second train leaves Gradel.
It travels towards Farham at a constant speed of 40 miles per hour.
d) On a copy of the graph, draw the journey of the second train.
e) At what time do the two trains pass each other?

3 Tess owns a car hire company.
The cost of hiring a car is £60 for the first day and then £36 per day after that.
a) Copy and complete this table to show the cost in £(C) to hire a car from Tess's car hire company by the end of each of the first four days.

Day number (n)	Total cost in £ (C)
1	
2	
3	
4	

b) Construct a graph to show the total cost by the end of each of the first 10 days.
c) Add a second line to your graph to show the cost if Tess had charged the same amount of £40 per day from the start.
d) At the end of which day would the cost be the same for both types of charges?
e) Write down two formulae for C, in terms of n, for the cost of hiring a car for n days using the two charging methods.

4 A group of hikers are hiking from their camp to a ruined castle.
They set off from their camp at 0800.
They walk for 2 hours at 3 km per hour.
They stop for a 30 minute rest.
After the rest, they walk on for a further $2\frac{1}{2}$ hours at 4 km per hour.
Then they stop for lunch for 1 hour.
After lunch, they walk on for a further 2 hours at 4 km per hour until they reach the ruined castle.

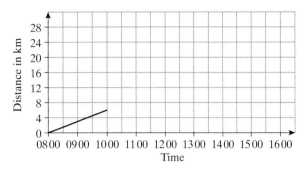

a) On a copy of this grid, complete the travel graph.
b) How far is the ruined castle from the camp?

✗ EXERCISE 11.3

1 Leo goes out in his car. He accelerates at a constant rate for the first 30 seconds, then travels at a steady velocity for the next 2 minutes. He then sees a pedestrian crossing ahead, so he slows down at a constant rate until he sees the pedestrian has crossed the road, before continuing at a steady velocity again.
The diagram below shows these parts of his journey, with corresponding line segments AB, BC, CD and DE.

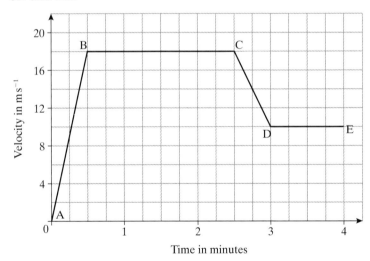

a) Work out Leo's acceleration during the first 30 seconds of his journey.
b) Write down the maximum velocity during his journey.
c) Work out how far Leo travelled during the section of the journey marked BC on the diagram.
d) Did Leo stop at the pedestrian crossing?

2 A spacecraft accelerates from rest to a speed of 3200 m s^{-1} in 3 minutes.
a) Convert 3 minutes into seconds.
b) Work out the acceleration of the spacecraft, in m s^{-2}.

3 A particle accelerates and then maintains a constant velocity before it decelerates and comes to rest again.

The velocity–time graph shows the movement of the particle.

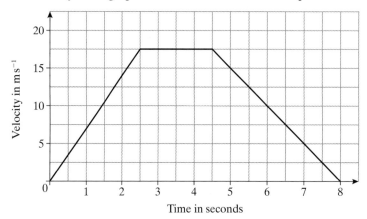

Time in seconds

a) Work out the acceleration of the particle during the first $2\frac{1}{2}$ seconds.
b) Work out the distance the particle travels when it maintains a steady velocity.
c) Work out the deceleration of the particle.

4 A motorbike is travelling at a constant velocity of $20\,\mathrm{m\,s^{-1}}$ from point A.
It continues at this velocity for 12 seconds until it reaches point B.
It then decelerates uniformly over the next 3 seconds to $15\,\mathrm{m\,s^{-1}}$ to point C.
It continues at this new velocity for 18 seconds, to reach point D.

a) Illustrate this information on a velocity–time graph.
Indicate the points A, B, C and D on your graph.

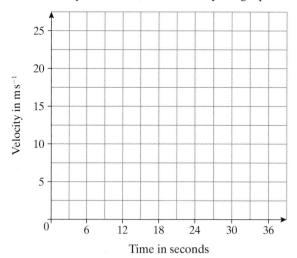

Time in seconds

b) Work out how far the motorbike travels between points A and B.
c) Work out how far the motorbike travels between points C and D.

CHAPTER 12

Working with shape and space

EXERCISE 12.1

X

Find the values of the angles represented by the letters in each diagram.

1

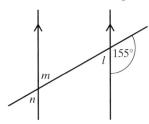

2

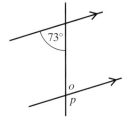

3

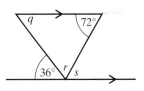

4

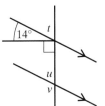

5

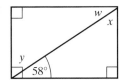

6

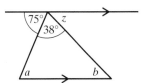

7

8

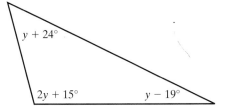

EXERCISE 12.2 **pt**

X

1 Find the value of *y*.

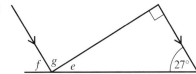

2 Find the value of *m*.

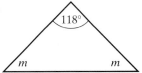

3

a) Form an equation in *x*.
b) Hence work out the size of the largest angle.

4

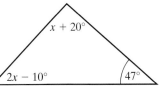

a) Form and solve an equation in *y*.
b) Hence find the sizes of the angles in the triangle.

5 A triangle has angles $3x + 16°$, $2x - 26°$ and $90°$.
  a) Set up an equation in x.
  b) Solve your equation to find the value of x.
  c) Work out the sizes of the angles in the triangle.

6 The angles in a triangle are $2y + 10°$, $6y - 32°$ and $4y + 10°$.
  a) Set up an equation in y.
  b) Solve your equation to find the value of y.
  c) Work out the sizes of the angles in the triangle.

7 The diagram shows a quadrilateral.
Work out the value of g.

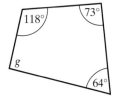

8 The diagram shows a quadrilateral.
Work out the value of y.

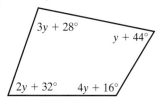

9 Find the size of each angle marked h.

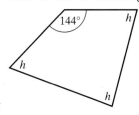

10 a) Form an equation in y.
  b) Solve your equation to find y.
  c) Hence find the size of each angle
in the quadrilateral.

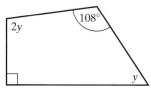

11 a) Form an equation in k.
  b) Solve your equation to find k.
  c) Hence find the size of each of the two angles on the straight line.

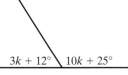

12 a) Set up an equation in x.
  b) Solve your equation to find the value of x.
  c) Work out the size of each angle.
  d) Check that your four angles add up to $360°$.

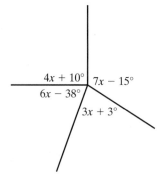

1 Find the sum of the interior angles of
   a) a hexagon   b) a 10-sided polygon.

2 Work out the size of the exterior angle at each vertex of
   a) a regular pentagon   b) a regular 12-sided polygon.

3 Seven of the angles in an octagon are 152°, 123°, 168°, 145°, 110°, 125° and 130°.
   Find the eighth angle.

4 a) The diagram shows part of a
       regular polygon.

   Work out how many sides the polygon has.

   b) Emma draws this diagram. She says it
       shows part of a regular polygon.

   Explain how you can tell that
   Emma must have made a mistake.

5 The diagram shows an irregular hexagon.
   Work out the value of the angle marked h.

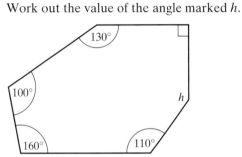

6 a) Work out the value of x.
   b) Hence work out the size of each angle.

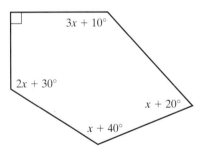

7 The diagram shows an octagon.
   All the angles marked w are equal.
   Calculate the value of w.

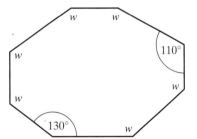

8 The diagram shows a pentagon. It has a vertical line of symmetry.
   a) Explain carefully why $a + b = 215°$.
       Angle a is 45° smaller than angle b.
   b) Use this information to rewrite the equation
       $a + b = 215°$ so it does not contain a.
       Solve this equation to find the value of b.
   c) Hence find the other angles in the pentagon.
   d) Julian says the pentagon is regular. Is he right? Explain your answer.

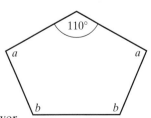

9 Follow these instructions to make an accurate drawing of a regular pentagon.

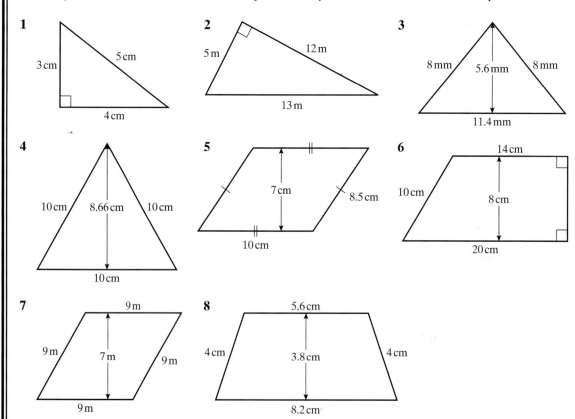

Step 1 Using compasses, construct a circle of radius 6 cm.

Step 2 Using your protractor, draw a sector using an angle of 72°.

Step 4 Complete the construction by joining the five points around the circumference of the circle.

Step 3 Repeat Step 2 to create a full set of 72° sectors.

72°

10 Adapt the instructions from question 9 to make:
   a) a regular hexagon
   b) a regular 10-sided polygon
   c) a 10-pointed star.

EXERCISE 12.4

Find the perimeter and the area of each shape. You may use standard formulae to help.

1 (triangle) 3 cm, 5 cm, 4 cm

2 (triangle) 5 m, 12 m, 13 m

3 (triangle) 8 mm, 5.6 mm, 8 mm, 11.4 mm

4 (triangle) 10 cm, 8.66 cm, 10 cm, 10 cm

5 (parallelogram) 7 cm, 8.5 cm, 10 cm

6 (trapezium) 14 cm, 10 cm, 8 cm, 20 cm

7 (parallelogram) 9 m, 9 m, 7 m, 9 m, 9 m

8 (trapezium) 5.6 cm, 4 cm, 3.8 cm, 4 cm, 8.2 cm

Calculate the perimeter and area of each shape. State the units in each case.

9

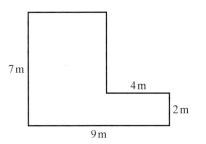

10

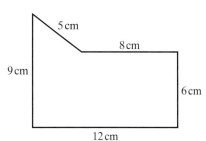

11

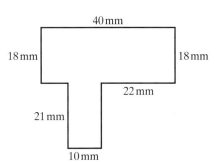

12

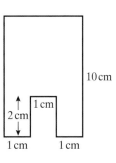

13

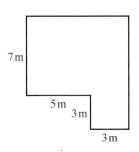

14

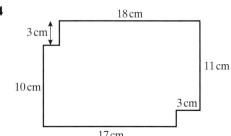

15 A quadrilateral has two pairs of parallel sides, but not all angles are the same size.
What type of quadrilateral is it?

16 A quadrilateral has all four angles the same size. Daksha says, 'It must be a rectangle.'
Is Daksha right? Explain your answer.

17 The diagram shows a rectangle.
The perimeter of the rectangle is 60 cm.
Work out the value of x.

18 The diagram shows an isosceles triangle. $AB = AC$.
All the lengths are in cm.
  a) Work out the value of x.
  b) Hence work out the perimeter of the triangle.

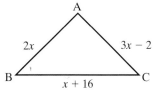

19 The diagram shows a triangle.
All the lengths are in cm.
a) What type of triangle is this?
b) Set up, and solve, an equation in x.
c) Hence work out the perimeter of the triangle.

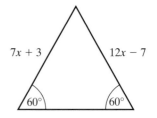

20 The diagram shows a parallelogram.
All lengths are in metres.
a) Set up, and solve, an equation in x.
b) Set up, and solve, an equation in y.
c) Hence work out the lengths of the sides
of the parallelogram.

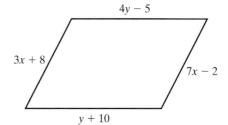

EXERCISE 12.5

1 The diagram shows a cube of side 3 m.
Calculate: a)   its surface area
                    b)   its volume.
State the units in your answers.

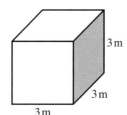

2 The diagram shows a cuboid with
dimensions 5 cm, 8 cm and 10 cm.
Calculate its   a)   surface area
                          b)   volume.
State the units in your answers.

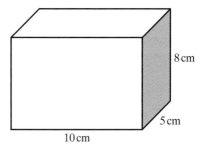

3 The diagram shows a prism.
Its cross section is formed by a right-angled triangle
of sides 6 cm, 8 cm and 10 cm.
The prism has a length of 12 cm.
a) Calculate the area of the cross section, shaded in
the diagram.
b) Work out the volume of the prism.
c) Calculate the surface area of the prism.

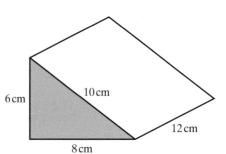

4 The cross section of a steel girder is in the shape of a letter T.
The cross section is shown in the diagram.
  a) Work out the area of the T-shaped cross section.

The girder is 50 cm long.
  b) Work out the volume of the girder.

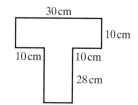

5 A cube measures 20 cm along each side.
  a) Work out the volume of the cube.　　**b)** Work out the surface area of the cube.

6 A cuboid measures 37 cm by 40 cm by 5 cm.
  a) Work out the volume of the cuboid.　　**b)** Work out the surface area of the cuboid.

7 The diagram shows a water tank.
It is in the shape of a cuboid.
It has no lid.
  a) Work out the volume of the tank, correct
     to 3 significant figures.
  b) Work out the total surface area of the
     inside of the tank.

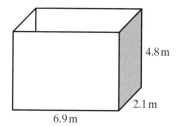

8 The diagram shows a sketch of a swimming pool.

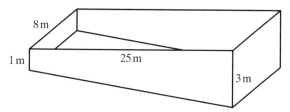

The pool is 1 m deep at the shallow end, and 3 m deep at the deep end.
The pool is 25 m long and 8 m wide.
  a) Work out the volume of the pool.
     1 cubic metre = 1000 litres
  b) Work out the number of litres of water in the pool when it is full.

9 A cube has a volume of 1000 cm³.
  a) Work out the dimensions of the cube.　　**b)** Calculate the surface area of the cube.

10 A cuboid has a volume of 40 cm³.
Its dimensions are all integers bigger than 1.
No two dimensions are the same.
  a) Work out the dimensions of the cuboid.　　**b)** Calculate the surface area of the cuboid.

CHAPTER 13

Circles and cylinders

 EXERCISE 13.1

1 A circle has radius 11 mm. Find its circumference, correct to 3 significant figures.

2 A circle has diameter 30 cm. Find its circumference, correct to 3 significant figures.

3 A circle has radius 15 cm. Find its area, correct to 3 significant figures.

4 A circle has diameter 42.8 cm. Find its area, correct to 3 significant figures.

5 Find, correct to 4 significant figures, the circumference of a circle with radius 37.18 cm.

6 Find, correct to 3 significant figures, the area of a circle with diameter 53.74 mm.

7 Find, correct to 4 significant figures, the circumference of a circle with diameter 3.06 cm.

8 Find, correct to 4 significant figures, the area of a circle with radius 0.924 cm.

Give the answers to each of these problems correct to 3 significant figures.

9 A circle has diameter 13 cm. Find its area.

10 A circle has radius 0.43 mm. Find its area.

11 A circle has diameter 370 cm. Find its circumference.

12 A circle has radius 8.05 m. Find its circumference.

13 A circle has a diameter of 24 mm. Calculate its circumference.

14 A face of a coin is a circle of diameter of 32.4 mm. Calculate its area.

15 Anna decides to run around a circular race track.
The radius of the track is 30 metres.
a) Work out the length of one lap of the track.

Anna wants to run at least 2000 metres.
She wants to run a whole number of laps.
b) Work out the minimum number of laps that Anna must run.

16 The diagram shows a rectangle inside a circle.
The circle has a radius of 50 cm.
The rectangle measures 80 cm by 60 cm.
The corners of the rectangle are on the circumference of the circle.
  a) Work out the area of the rectangle.
  b) Work out the area of the circle.
  c) Work out the shaded area.
     Give your answer correct to the nearest square centimetre.

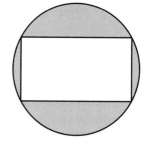

17 The diagram shows a rectangular piece of card
with four circles cut out of it.
The circles are each of diameter 5 cm.
The rectangle measures 24 cm by 13 cm.
  a) Calculate the area of one of the circular holes.
  b) Work out the area of the card.

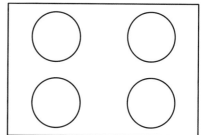

18 A circular table stands on a circular base.
The diameter of the table top is 94 cm.
The radius of the base is 33 cm.
  a) Calculate the area of the table top.
  b) Calculate the circumference of the base.

19 A bicycle wheel has a diameter of 70 cm.
  a) How far, in metres, does the wheel go when it does one complete revolution?
  b) How far, in metres, does the wheel go when it does 20 complete revolutions?
  c) How many revolutions does the wheel do when it travels it 80 m?

20 The diagram shows a circular garden with a circular pond
in the middle.
The whole garden, apart from the pond, is lawn.
The garden has a radius of 5 m.
The pond has a diameter of 2 m.
  a) Calculate the area of the pond.
  b) Calculate the area of the garden.
  c) Hence find the area of the lawn.

The gardener wants to put some fertiliser on the lawn.
1 bag of fertiliser is sufficient for 0.8 m².
  d) Work out how many bags of fertiliser the gardener
     needs to buy.

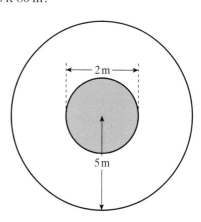

21 The diagram shows a running track at a school.
It is made up of two straight sections, and two
semicircular ends.
The dimensions are marked on the diagram.
Andy runs around the outside boundary of the
track, marked with a solid line.
Colin runs around the inside boundary, marked
with a dotted line.
They each run one lap of the track.

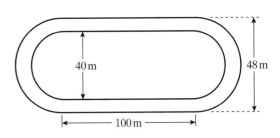

a) Work out how far Colin runs.
b) Work out how much further than Colin, Andy runs.

EXERCISE 13.2 🔵

Calculate the perimeter and area of each sector. Give each answer correct to 3 significant figures.

1

8 cm
39°
8 cm

2
5 cm
76°
5 cm

3
6.7 cm
135°
6.7 cm

4
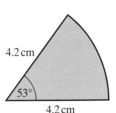
4.2 cm
53°
4.2 cm

5
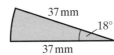
37 mm
18°
37 mm

6
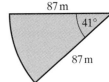
87 m
41°
87 m

7
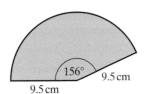
156°
9.5 cm
9.5 cm

8
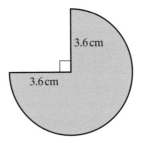
3.6 cm
3.6 cm

9 Bob displays some statistical data in a pie chart.
The three sectors of the pie chart are to have angles of 210°, 50° and 100°.
The radius of the pie chart is 7 cm.
a) Work out the area of each sector of Bob's pie chart.
b) Work out the perimeter of the smallest sector.

10 A pizza of diameter 18 inches is to be shared between 10 people. It is cut into ten equal sectors.
a) Work out the angles at the centre of each sector of pizza.
b) Work out the area of one sector.

EXERCISE 13.3

Give the answers to each of these problems correct to 3 significant figures.

1 A circle has circumference 28 mm. Find its radius.

2 A circle has circumference 38.2 cm. Find its diameter.

3 A circle has area 120 cm². Find its radius.

4 A circle has area 60 cm². Find its radius.

5 A circle has circumference 74 m. Find its diameter.

6 A circle has area 850 cm². Find its diameter.

7 A circle has circumference 7.9 cm. Find its diameter.

8 A circle has area 483 cm². Find its radius.

9 The diagram shows a sector of a circle.
The radius of the sector is r centimetres.
The area of the sector is 26.1 cm².
  a) Work out the area of the corresponding complete circle.
  b) Hence find the value of r.
  c) Calculate the perimeter of the sector.

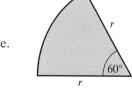

10 The diagram shows a running track.
The ends are semicircles of radius y metres.
The straights are of length 45 metres each.
The total distance around the track is 120 metres.
  a) Calculate the value of y.
  b) Calculate the area contained within the
    running track.

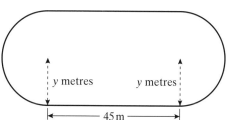

EXERCISE 13.4

Give the answers to each of these problems correct to 3 significant figures.

1 A cylinder has radius 15 cm and height 23 cm. Find its volume.

2 A cylinder has radius 4 cm and height 3 cm. Find its curved surface area.

3 A cylinder has diameter 36 cm and height 10 cm.
  a) Find its volume.
  b) Find its curved surface area.

4 A cylinder has radius 7 cm and height 18 cm. Find its volume.

5 The diagram shows a hollow cylinder.
Work out the curved surface area of the cylinder.

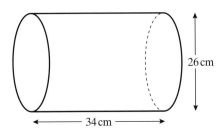

26 cm

34 cm

6 A hollow cylindrical pipeline has an internal diameter of 16 cm.
The pipeline is 250 metres in length.
a) Work out the volume of the pipeline.
Give your answer in cm^3.
b) 1000 cm^3 = 1 litre
Express the volume of the pipeline in litres.

7 A tube of crisps is in the shape of a cylinder.
It has radius 7 cm and height 30 cm.
Work out the volume of the tube.

8 A sweet packet is in the shape of a hollow cardboard cylinder.
The inside diameter of the cylinder is 4.5 cm and it has a height of 12 cm.
a) Work out the volume of the cylinder.

The sweets have a volume of 2.3 cm^3 each.
b) Show that the packet cannot contain as many as 90 sweets.

9 Rosie wants to buy a can of soup.
She wants to buy the can of soup that has the greatest volume.

A

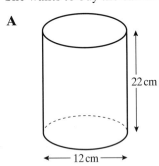

22 cm

12 cm

B

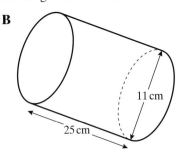

11 cm

25 cm

Which of these two cans should she buy?
Show all your working.

10 A cylinder of radius 6.2 cm has a volume of 1183 cm^3, correct to 4 significant figures.
Work out the height of the cylinder.

X EXERCISE 13.5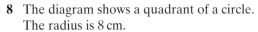

1 A circle has diameter 30 cm. Work out:
   a) its circumference   **b)** its area.
   Leave your answers in terms of π.

2 A circle has radius 18 cm. Work out:
   a) its circumference   **b)** its area.
   Leave your answers in terms of π.

3 A cylinder has radius 10 cm and height 7 cm. Work out:
   a) its curved surface area   **b)** its volume.
   Leave your answers in terms of π.

4 A circle has circumference 39π cm.
   a) Work out the exact radius of the circle.
   b) Work out the exact area of the circle.
   Leave your answers in terms of π.

5 A circle has a circumference of 5π cm.
   a) Work out the exact radius of the circle.
   b) Work out the exact area of the circle.
   Leave your answers in terms of π.

6 A cylinder has volume 580π cm³.
   Its radius is 20 cm.
   Work out its height.

7 A cylinder has volume 360π cm³.
   Its radius is 10 cm.
   Work out its height.

8 The diagram shows a quadrant of a circle.
   The radius is 8 cm.
   a) Work out the area of the quadrant, in terms of π.
   b) Work out the perimeter of the quadrant.
      Give your answer correct to 3 significant figures.

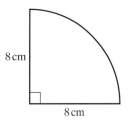

9 The diagram shows an unusual carpet.
   It is in the shape of a square, with semicircles on each of the four sides.
   Each side of the square is 3.4 m.
   a) Work out the area of one of the semicircles.
      Leave your answer in terms of π.
   b) Hence find the total area of the carpet.
      Give your answer correct to 3 significant figures.

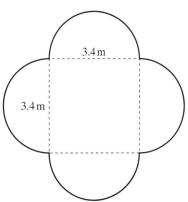

10 The diagram shows two cylinders.
   Cylinder A has diameter 5 cm and height 12 cm.
   Cylinder B has diameter 12 cm and height 5 cm.
   a) Show that both cylinders have exactly
      the same curved surface area.
   b) Work out the volume of each cylinder,
      leaving your answers in terms of π.
   c) Which cylinder has the larger volume?

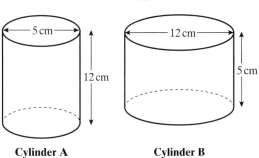

Cylinder A   Cylinder B

CHAPTER 14

Constructions and loci

EXERCISE 14.1

Make accurate drawings of these triangles, stating which of the various combinations of information you have been given: SAS, ASA, SSS or SSA. If any triangles are ambiguous, draw both possibilities.

1 Draw triangle ABC with BC = 9 cm, AC = 7 cm, angle BCA = 72°.

2 Draw triangle PQR with PQ = 8 cm, angle RPQ = 43°, angle RQP = 70°.

3 Draw triangle KLM with KL = 5 cm, LM = 4 cm, angle KLM = 150°.

4 Draw triangle DEF with DE = 9 cm, EF = 5.5 cm, angle DEF = 58°.

5 Draw triangle EFG with EF = 4.2 cm, FG = 7.5 cm, EG = 10 cm.

6 Draw triangle JKL with JK = 8.1 cm, JL = 5 cm, LK = 5 cm.

7 The sketch shows a triangle with
AB = 73 mm, BC = 56 mm, AC = 87 mm.
Make an accurate drawing of this triangle.

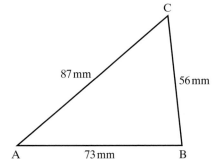

8 The sketch shows a triangle with AB = 84 mm,
BC = 52 mm and angle ACB = 90°.
Make an accurate drawing of this triangle.

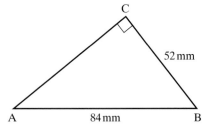

9 Draw triangle PQR with PQ = 8.1 cm, QR = 8.8 cm, angle RPQ = 90°.
Use ruler and compasses to construct this triangle.
Show that there are two different solutions based on the given information.

10 Using compasses, try to make an accurate construction of triangle EFG with sides
EF = 14 cm, FG = 6 cm, EG = 7 cm. What difficulty do you encounter?
Find a reason for this.

EXERCISE 14.2

1 Use ruler and compasses to construct the perpendicular bisector of the line segment AB
(AB is 7 cm long).

A ——————————————— B

2 Use ruler and compasses to construct the bisector of angle ABC.

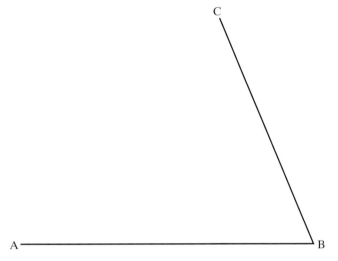

3 Use ruler and compasses to construct the perpendicular from A to the line segment PQ.

×A

P ——————————————— Q

4 Use ruler and compasses to construct a line passing through X, perpendicular to the line AB.

A —————— ✕ —————— B

5 Use ruler and compasses to construct the bisector of angle CDE.

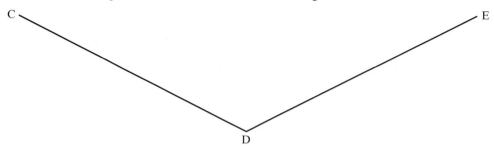

6 Use ruler and compasses to construct the perpendicular bisector of the line segment PQ (PQ is 9.6 cm long).

7 Use ruler and compasses to construct a line passing through ✗, perpendicular to the line CD.

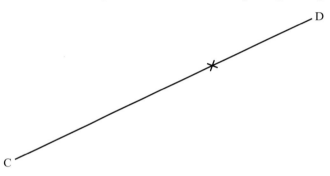

8 Use ruler and compasses to construct the perpendicular from P to the line segment AB.

✗ P

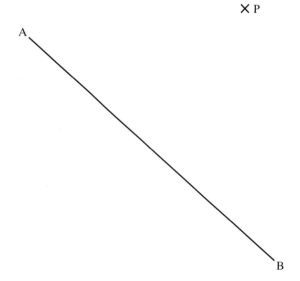

14 Constructions and loci

EXERCISE 14.3

1 In the diagram, the rectangle PQRS represents a farm. It is to be divided up into three regions.

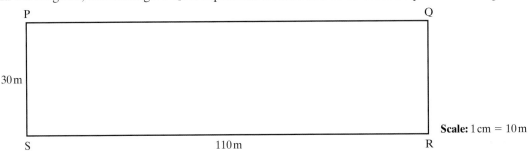

The region formed by all points within 30 metres of PS will be the farm house.
The region formed by all points within 40 metres of Q will be the hen house.
The rest of the farm will be covered with grass.
On the copy of the diagram, indicate each of the three regions.

2 The diagram shows a paved play area at a school.

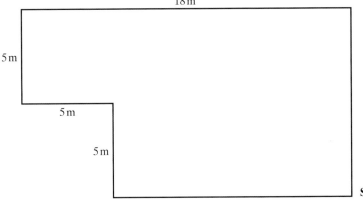

The local authority wants to erect a fence around the outside of the play area.
They decide that the fence should come to within 2 metres of the nearest point of the play area.
On a copy of the diagram, make a drawing to show the position of the fence.

3 The diagram shows a sketch of a rectangular nature reserve.
A gravel path is to be laid.
The centre line of the path runs diagonally from E to G.
All parts of the nature reserve within 1 metre of this centre
line are to be gravelled.

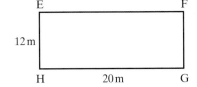

   a) Using a scale of 1 cm = 2 m, draw the rectangle.
   b) Mark the centre line of the path.
   c) Shade the region to be gravelled.

A pond is to be built in the nature reserve.
The pond will occupy all points within 4.6 m of F.
   d) Shade the area where the pond will be.

4 At the edge of a bay there are three lighthouses.
The diagram shows the position of the three lighthouses, P, Q and R.

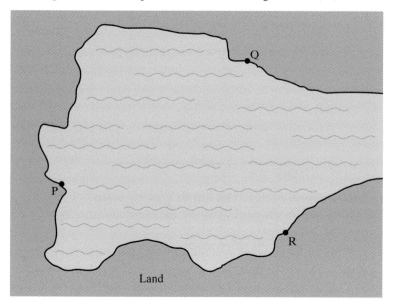

Land

a) Using compasses and a straight edge, construct the locus of all points that are equidistant from P and Q.

b) Repeat the construction using Q and R, and again, using P and R.

c) Hence divide the bay into three regions, one for each lighthouse.
Use coloured pencils to mark the regions distinctly.

5 The diagram shows a sketch of a rectangular field ABCD.
Harry wants to bury some treasure in the field.

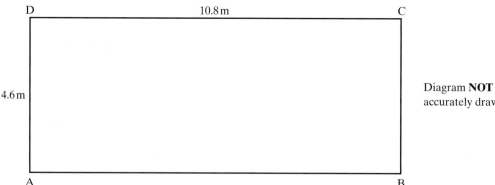

D   10.8 m   C

4.6 m

Diagram **NOT** accurately drawn

A   B

a) Using a scale of 1 cm = 1 m, construct the rectangular field ABCD.

Harry buries the treasure exactly 3.5 m from DC, and exactly the same distance from AB and BC.

b) On a copy of the diagram, mark with a cross (✗) where Harry buried the treasure.

CHAPTER 15

Transformation and similarity

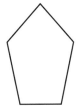

EXERCISE 15.1

1 Diagram **a)** shows a 2-D shape. Copy the diagram and draw a line of symmetry on your diagram.

Diagram **b)** shows a sketch of a 3-D object. Copy the diagram and indicate a plane of symmetry on your sketch.

a)   **b)**

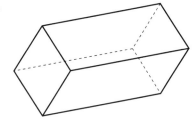

In questions **2** to **5**, copy the diagram and draw the reflection of the given shape in the mirror line indicated. Label the mirror line with its equation in each case.

2

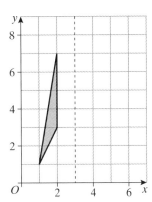

3

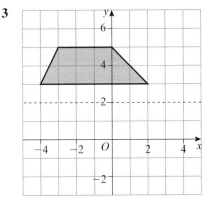

4

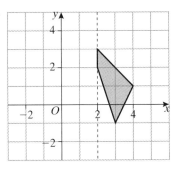

5

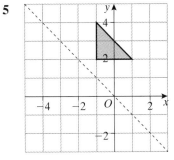

6 The diagram shows a triangle P and its mirror
image Q.
  a) Copy the diagram and draw the mirror line
    that has been used for the reflection.
  b) Write down the equation of the mirror line.

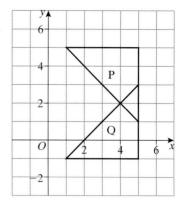

7 The diagram shows a triangle S and its mirror image T.
  a) Copy the diagram and draw the mirror line
    that has been used for the reflection.
  b) Write down the equation of the mirror line.

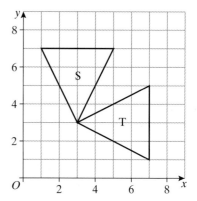

8 The diagram shows a letter T shape, labelled X. The shape is to be reflected in a mirror line.
  Part of the reflection has been drawn on the diagram.

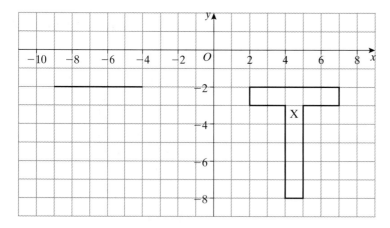

  a) Copy the diagram and complete the drawing to show the image. Label it Y.
  b) Mark the mirror line and give its equation.

9 The diagram shows ten triangles A to J.
The ten triangles are all congruent to
each other.

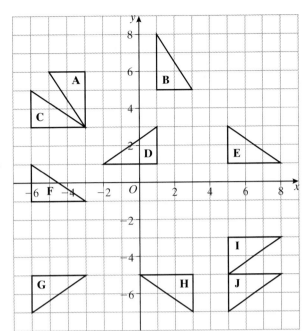

a) Explain the meaning of the word
   congruent.

b) Triangle E can be reflected to
   triangle I. State the equation of the
   mirror line that achieves this.

c) Triangle H is reflected to another
   triangle using the mirror line $x = -1\frac{1}{2}$.
   Which one?

d) Triangle D can be reflected to
   triangle E using a mirror line.
   Give the equation of this line.

e) Triangle B can be reflected to
   triangle E using a mirror line.
   Give the equation of this line.

f) Triangle A can be reflected to
   triangle C using a mirror line.
   Give the equation of this line.

g) Triangle G is reflected to another
   triangle using the mirror line $y = -3$. Which one?

10 A triangle P is reflected in the line $x = 3$ to form an image, triangle Q.
Then triangle Q is reflected in the same mirror line to form an image, triangle R.
What can you deduce about triangle P and triangle R?

EXERCISE 15.2

For each of questions **1** to **6**, copy the diagram on to a coordinate grid on which x and y range from
-8 to 8.

1 Rotate the trapezium shape
   $90°$ clockwise about O.

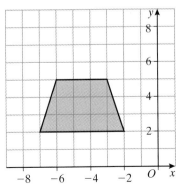

2 Rotate the L-shape $180°$ about O.

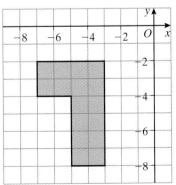

3

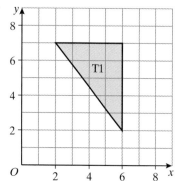

a) Rotate the triangle T1 90° anticlockwise about O.
   Label the result T2.
b) Rotate T2 180° about O.
   Label the result T3.
c) Describe the single rotation that takes T1 directly to T3.

4

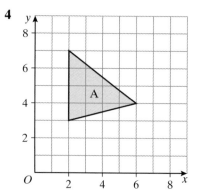

a) Rotate triangle A 90° anticlockwise about $(0, 0)$.
   Label the result B.
b) Rotate triangle B by 180° about $(0, 0)$.
   Label the result C.
c) Describe carefully the single rotation that takes triangle C to triangle A.

5

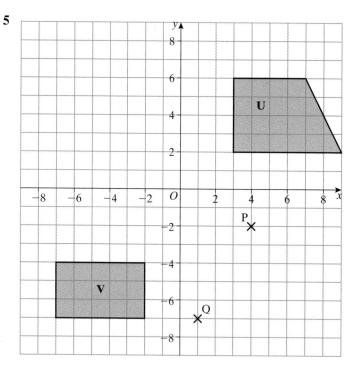

a) Rotate shape U 90° anticlockwise about point P $(4, -2)$.
b) Rotate shape V 90° clockwise about point Q $(1, -7)$.

15 Transformation and similarity

6 a) Rotate triangle A 90° clockwise about (1, 1). Label it B.
b) Now rotate both triangle A and triangle B by 180° about (1, 1).

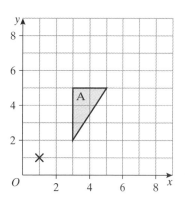

7 The diagram shows an object A and its image B after a rotation.
a) Write down the size and direction of the angle of rotation.
b) Write down the coordinates of the centre of rotation.

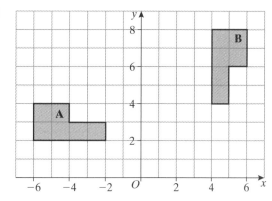

8 Mel says, 'If you translate a shape and then translate it again, the result is equivalent to a single translation.'

Joe says, 'If you reflect a shape and then reflect it again, the result is equivalent to a single reflection.'

Emily says, 'I'm afraid only one of you is right.'

Who is right, and who is wrong?

EXERCISE 15.3

For each of questions **1** to **6**, copy the diagram on to a coordinate grid on which x and y range from −8 to 8.

1 a) Reflect triangle S in the line $x = -1$.
Label the new triangle T.
b) Reflect triangle T in the x axis.
Label the new triangle U.
c) Describe the **single** transformation that maps S to U.

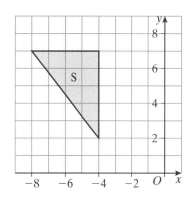

2 The diagram shows a triangle, T.

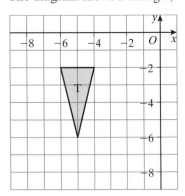

a) Translate triangle T by $\begin{pmatrix} 0 \\ 4 \end{pmatrix}$.

Label its image U.

b) Rotate triangle U by 90°
clockwise about O.
Label the resulting triangle V.

c) Describe the single transformation
that maps T to V.

3 The diagram shows a triangle, S.

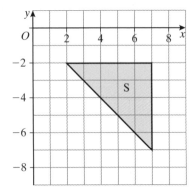

a) Reflect triangle S in the y axis.
Label its image T.

b) Reflect triangle T in the line
$y = -1$.
Label its image U.

c) Describe the single
transformation that maps S
directly to U.

4 The diagram shows a letter F shape.
The shape is labelled F1.

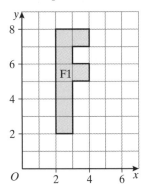

a) Reflect F1 in the y axis.
Label the image F2.

b) Reflect F2 in the line $y = x$.
Label this image F3.

c) Describe the single
transformation that would take
F3 to F1.

5 The diagram shows a quadrilateral E.

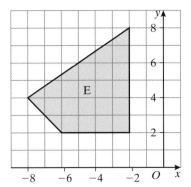

a) Rotate quadrilateral E through
90° anticlockwise about O.
Label the result F.

b) Rotate quadrilateral F through
90° clockwise about $(1, -1)$.
Label the result G.

c) Describe a single transformation
that would take G to E.

6 The diagram shows a triangle P.
   a) Rotate triangle P 90° anticlockwise about (0, 1).
      Label this result Q.
   b) Translate triangle P by the vector $\begin{pmatrix} 9 \\ 5 \end{pmatrix}$.

      Label this result R.
   c) Describe the single transformation that transforms
      triangle R to triangle Q.

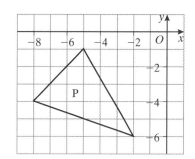

EXERCISE 15.4

1 Copy the diagram showing shape A and point P on to a
   coordinate grid on which x and y range from 0 to 18.
   a) Enlarge shape A by scale factor 2, centre P.
      Label the new shape B.
   b) Enlarge shape A by scale factor 3, centre P.
      Label the new shape C.
   c) Are shapes B and C:
      (i) congruent   **(ii)** similar?

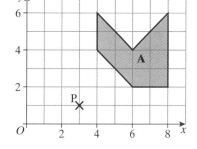

2 Copy the triangle, and a centre of enlargement P, on to a
   coordinate grid on which x ranges from 0 to 18 and y ranges
   from 0 to 10.

   Enlarge the triangle by scale factor −2, centre P.

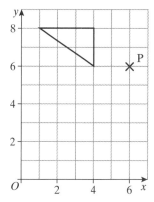

3 Draw the trapezium on to a coordinate grid on which x and y
   range from −10 to 10.
   Add centre of enlargement P at (6, −8).
   Enlarge the shape by scale factor $2\frac{1}{2}$, centred on P.

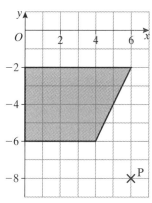

4 On a coordinate grid on which x ranges from 0 to 26 and y ranges from 0 to 16, draw shape P and centres A and B.

a) Enlarge shape P, with scale factor 3, centre A (2, 4). Label the result Q.

b) Enlarge shape Q, with scale factor $\frac{1}{3}$, centre B (14, −2). Label the result R.

c) State whether shapes P and Q are
  (i) congruent   **(ii)** similar.

d) Are shapes P and R congruent?

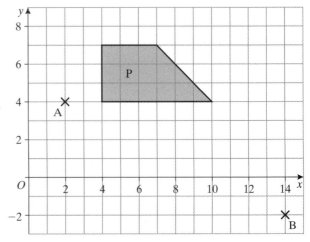

5 The diagram shows an object, A, and its image B after an enlargement.

a) State the scale factor for the enlargement.

b) Obtain the coordinates of the centre of enlargement.

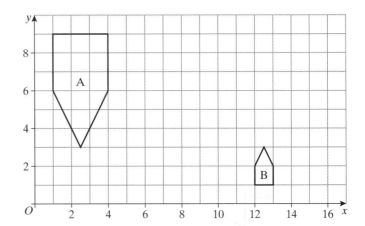

6 Ned draws a triangle and then enlarges it.
He notices that the image of the triangle is congruent to the original triangle.

a) What scale factor do you think Ned has used?

b) What other transformation could Ned have used to get the same result?

EXERCISE 15.5

1 The diagram shows two triangles.
They are mathematically similar.

a) Work out the height of the larger triangle.

b) Work out the ratio of the perimeters of the triangles, in the form 1 : n.

c) Find the ratio of the areas of the triangles, in the form 1 : n.

2 The diagram shows two similar parallelograms.
The smaller parallelogram has an area of 34 cm².
Work out the area of the larger parallelogram.

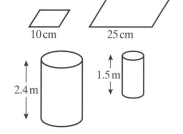

10 cm   25 cm

3 The diagram shows two solid cylinders.
They are similar.
Both cylinders are made of the same material.
The larger cylinder has a mass of 24.576 kg.
Work out the mass of the smaller cylinder.

2.4 m   1.5 m

4 In the diagram, AB and CD are parallel.

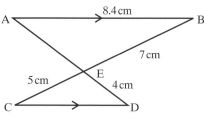

8.4 cm
A ————————————→ B
5 cm   E
          7 cm
          4 cm
C ————————————→ D

a) **(i)** Work out the length of AE.
   (ii) Hence find the length of AD.
b) Calculate the length of CD.

5 A shop has two dolls' houses.
They are mathematically similar.
One of the dolls' houses is 50% larger than the other one.
The smaller dolls' house has a length of 76 cm and a mass of 6 kg.
a) Calculate the length of the larger dolls' house.
The two dolls' houses are made from the same materials.
b) Calculate the mass of the larger dolls' house.

6 In the diagram, PQ and ST are parallel.
Calculate the lengths x and y.

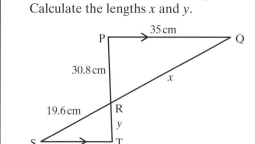

35 cm
P ————→ Q
30.8 cm
          x
19.6 cm   R
          y
S ————→ T
  10 cm

7 Bill makes two similar garden gnomes.
The smaller one is 24 cm tall and weighs 1.6 kg.
The larger one is 40 cm tall.
Work out the weight of the larger garden gnome.

8 The diagram shows triangle ABC and triangle ADE.
The line segments DE and BC are parallel.
AD = 10 cm, AE = 9 cm, DE = 15 cm, BC = 20 cm.
a) Explain fully why triangles ABC and ADE are similar.
b) Work out the length AC.   **c)** Work out the length BD.

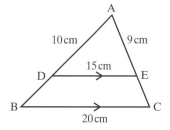

          A
10 cm   9 cm
     15 cm
D ————————→ E
B ————————————→ C
     20 cm

9 An architect builds a large model of a new school building she has designed.
The height of the model is 0.6 m and the model covers an area of 5 m².
The new school building will be mathematically similar to the model but 20 times larger.
Calculate: **a)** the height of the new school building; **b)** the area the new school will cover.

10 Two buckets are mathematically similar.
The height of the smaller bucket is 30 cm and it can hold 4 litres of water.
The larger bucket holds 8 litres of water.
Calculate the height of the larger bucket. Give your answer correct to 1 decimal place.

CHAPTER 16

Pythagoras' theorem

 EXERCISE 16.1

Look at these triangles, and use Pythagoras' theorem to decide whether they are right angled or not. The diagrams are not drawn to scale.

1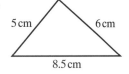
5 cm   6 cm
8.5 cm

2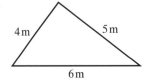
4 m   5 m
6 m

3   15 mm
9 mm   12 mm

For each of the triangles described below, use Pythagoras' theorem to decide whether it is right angled. If so, name the angle at which the right angle is located.

4 AB = 9 cm, BC = 7 cm, CA = 5.4 cm.   **5** AB = 2.5 cm, BC = 1.5 cm, CA = 2 cm.

6 AB = 8 mm, BC = 8 mm, CA = 11 mm.   **7** PQ = 2.6 cm, QR = 3.5 cm, RS = 5 cm.

8 PQ = 2.5 m, QR = 6 m, RS = 6.5 m.   **9** PQ = 7.2 cm, QR = 8.3 cm, RS = 12 cm.

10 AB = 15 km, BC = 20 km, CA = 25 km.

 EXERCISE 16.2

In questions **1** to **9**, find the length of the hypotenuse represented by the letters *a* to *i*. Give your answers to 3 significant figures where appropriate.

1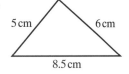
3 cm   *a*
7 cm

2
b   2 cm
8 cm

3
5.2 m   6.3 m
c

4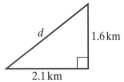
d   1.6 km
2.1 km

5
4.7 mm   *e*
2.8 mm

6
f   9.3 cm
6.7 cm

7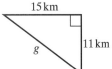
15 km
11 km
g

8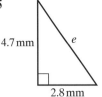
h
5 mm   8.7 mm

9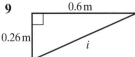
0.6 m
0.26 m   *i*

Find the length of the diagonal of each rectangle.

10

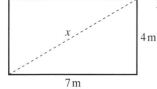

x
4 m
7 m

11

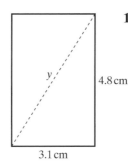

y
4.8 cm
3.1 cm

12

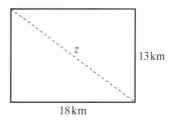

z
13 km
18 km

EXERCISE 16.3

Find the length of the side marked by the letters a to i below.
Give your answers to 3 significant figures where appropriate.

1

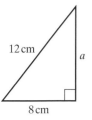

12 cm
a
8 cm

2

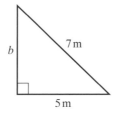

b
7 m
5 m

3

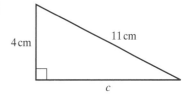

4 cm
11 cm
c

4

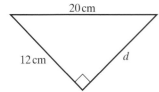

20 cm
12 cm
d

5

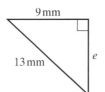

9 mm
13 mm
e

6

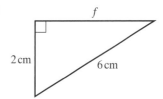

f
2 cm
6 cm

7

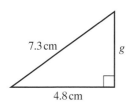

7.3 cm
g
4.8 cm

8

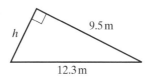

h
9.5 m
12.3 m

9

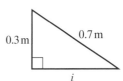

0.3 m
0.7 m
i

10 A rectangle has length y cm and width 6 cm. Its diagonal is of length 6.5 cm.
Find the value of y.

11 A ship sails due West for x km and then sails due South for 20 km.
It ends up 25 km in a direct straight line from its starting point.
Find the value of x.

The last part of this exercise contains a mixture of questions. Remember to square and add when you are finding a hypotenuse, but square and subtract when finding a shorter side.
In both cases, remember to square root at the end.

In questions **12** to **15** find the length of the side marked by the letters *a* to *d*.
Give your answers to 3 significant figures where appropriate.

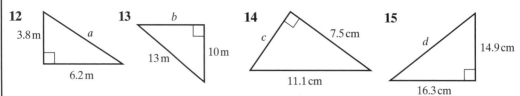

12   3.8 m   *a*   6.2 m

13   *b*   13 m   10 m

14   *c*   7.5 cm   11.1 cm

15   *d*   14.9 cm   16.3 cm

EXERCISE 16.4

Give all answers correct to 3 significant figures.

1 The diagram shows a box in the shape of a cuboid.
  a) Work out the length FH.
  b) Work out the length HB.

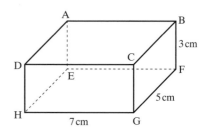

A   B   3 cm   C   D   E   F   5 cm   H   7 cm   G

2 The diagram shows a box in the shape of a cuboid.
  a) Work out the length BD.
  b) Work out the length BH.

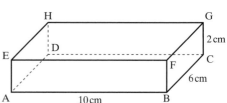

H   G   2 cm   D   E   C   F   6 cm   A   10 cm   B

3 A cuboid measures 3 cm by 5 cm by 9 cm.
  a) Make a sketch of the cuboid.
  b) Calculate the length of the diagonal.

4 A box measures 8 cm by 13 cm by 18 cm.
Rob has a giant pencil of length 25 cm.
Use Pythagoras' theorem to explain whether
it is possible to put Rob's pencil in the box
so that it does not stick out of the box.

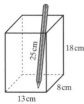

25 cm   18 cm   8 cm   13 cm

5 The diagram shows a wedge.
The rectangular face ABDE is at right angles
to the rectangular face CBEF.
DE = 14 cm, EF = 6 cm, BE = 20 cm.
  a) Calculate the length BD.
  b) Calculate the direct distance from C to D.

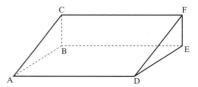

C   F   B   E   A   D

6 A thin rod of length 15 cm just
fits inside a cylindrical container
of length 11 cm.
Find the diameter of the cylinder.

11 cm   15 cm

7 Find the length of the longest
thin rod that will just fit inside
a cuboid-shaped box with
dimensions 6 cm by 8 cm by
26 cm.

CHAPTER 17

Introducing trigonometry

EXERCISE 17.1

Work out the values of the sides represented by letters.
Round your answers correct to 3 significant figures.

1
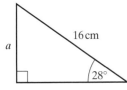
16 cm
a
28°

2
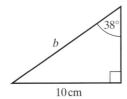
38°
b
10 cm

3
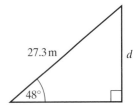
3 cm
c
72°

4
27.3 m
48°
d

5
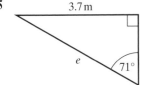
3.7 m
e
71°

6
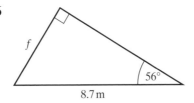
f
56°
8.7 m

7
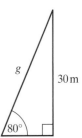
g
30 m
80°

8
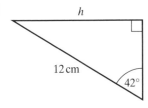
h
12 cm
42°

9
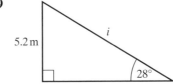
5.2 m
i
28°

10
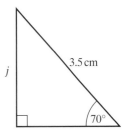
3.5 cm
j
70°

11 In triangle ABC, BC = 20 cm, angle ABC = 90°, angle CAB = 35°. Calculate AC.

12 In triangle DEF, angle DEF = 90°, angle DFE = 74°, DE = 8.6 mm. Calculate DF.

13 In triangle GHJ, angle GHJ = 90°, angle GJH = 50°, GJ = 83 cm. Calculate GH.

14 In triangle KLM, angle KLM = 90°, angle LMK = 12°, KL = 9.4 mm. Calculate KM.

15 In triangle NPQ, angle NPQ = 90°, angle PNQ = 69°, NQ = 293 m. Calculate PQ.

EXERCISE 17.2

Work out the values of the sides represented by letters.
Give your answers correct to 3 significant figures.

1

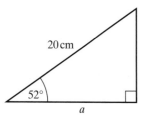

20 cm

52°

a

2

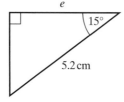

b

42°

12 cm

3

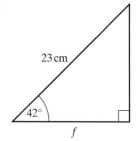

c

25°

5.6 cm

4

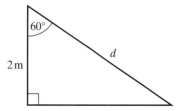

60°

2 m

d

5

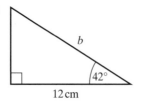

e

15°

5.2 cm

6

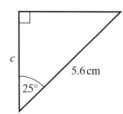

23 cm

42°

f

7

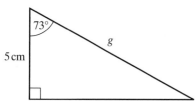

73°

5 cm

g

8

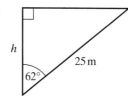

h

25 m

62°

9

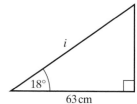

i

18°

63 cm

10

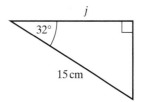

j

32°

15 cm

11 In triangle ABC, AC = 8.2 cm, angle ABC = 90°, angle ACB = 61°. Calculate BC.

12 In triangle DEF, angle DEF = 90°, angle FDE = 37°, DE = 9 cm. Calculate DF.

13 In triangle GHJ, angle GHJ = 90°, angle GJH = 56°, GJ = 125 m. Calculate HJ.

14 In triangle KLM, angle KLM = 90°, angle LKM = 23°, LK = 57 mm. Calculate KM.

15 In triangle NPQ, angle NPQ = 90°, angle PQN = 40°, NQ = 66.6 cm. Calculate PQ.

EXERCISE 17.3

Work out the values of the sides represented by letters.
Give your answers correct to 3 significant figures.

1
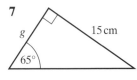
13°
21 cm
a

2
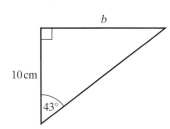
b
10 cm
43°

3
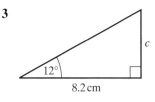
c
12°
8.2 cm

4
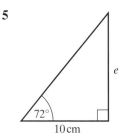
10.3 m
d
18°

5
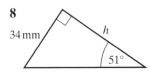
e
72°
10 cm

6
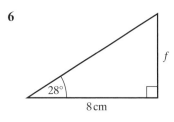
f
28°
8 cm

7
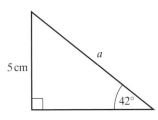
g
15 cm
65°

8
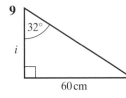
34 mm
h
51°

9

32°
i
60 cm

10
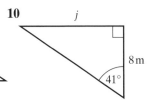
j
8 m
41°

11 In triangle ABC, BC = 20 cm, angle ABC = 90°, angle CAB = 23°. Calculate AB.

12 In triangle KLM, angle KLM = 90°, angle KML = 18°, KL = 11.4 cm. Calculate LM.

EXERCISE 17.4 pt

Work out the unknown lengths represented by letters.
Round your answers correct to 3 significant figures.

1
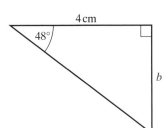
5 cm
a
42°

2
4 cm
48°
b

3
10 m
c
38°

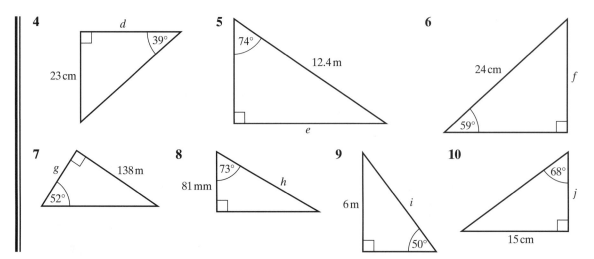

4 d, 23 cm, 39°

5 74°, 12.4 m, e

6 24 cm, f, 59°

7 g, 138 m, 52°

8 73°, 81 mm, h

9 6 m, i, 50°

10 68°, j, 15 cm

EXERCISE 17.5

Work out the values of the angles represented by letters.

Round your final answers correct 1 decimal place.

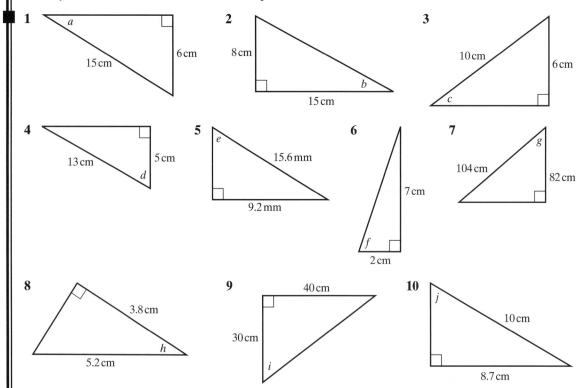

1 a, 15 cm, 6 cm

2 8 cm, b, 15 cm

3 10 cm, 6 cm, c

4 13 cm, 5 cm, d

5 e, 15.6 mm, 9.2 mm

6 f, 7 cm, 2 cm

7 g, 104 cm, 82 cm

8 3.8 cm, h, 5.2 cm

9 40 cm, 30 cm, i

10 j, 10 cm, 8.7 cm

11 In triangle ABC, angle ABC = 90°, AB = 18 cm, AC = 23 cm. Calculate angle ACB.

12 In triangle DEF, angle DEF = 90°, EF = 6.2 cm, DE = 4.3 cm. Calculate angle EFD.

EXERCISE 17.6

1 A ship sails 27 km due South, then 16 km due West.
 a) Illustrate this information on a sketch.
 b) How far, to the nearest 0.1 km, is the ship from its starting point?
 c) What bearing, to the nearest degree, should it steer to return directly to its start point?

2 In the diagram, PRS is a straight line. PQ = 9.2 cm,
 QS = 7.6 cm and angle QPR = 38°.
 QR is perpendicular to PS.
 a) Calculate the length QR.
  Give your answer to 3 significant figures.
 b) Calculate angle QSP.
  Give your answer to the nearest 0.1 cm.
 c) Calculate angle PQS.
  Give your answer to the nearest 0.1°.

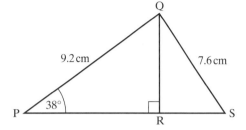

3 The diagram shows a cross section of a triangular prism.
 AB = BC = 7 cm. BM = 5 cm.
 M is the midpoint of the line segment AC.
 a) Explain why angle AMB must be a right angle.
 b) Calculate the angle BCM. Give your answer to the nearest 0.1°.
 c) Calculate the area of the triangle.
  Give your answer to 3 significant figures.

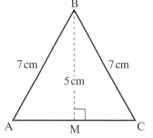

4 A 5 m ladder leans against a vertical wall. The bottom of the ladder is 2.8 m from the wall.
 a) Illustrate this information on a sketch.
 b) Calculate the angle between the ladder and the wall.
  Give your answer correct to 1 decimal place.
 c) Calculate how high the top of the ladder is above the ground.
  Give your answer to 3 significant figures.

5 In the diagram, EHG is a straight line. Angle EHF = 90°.
 Angle FGH = 68°. EG = 26 cm, FH = 15 cm.
 a) Work out the length of GH, correct to 3 significant figures.
 b) Work out the size of angle FEH, correct to the
  nearest degree.

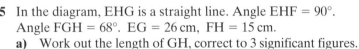

6 Annerly is due North of Carr and is 300 km away.
 Bronwich is between Annerly and Carr.
 Dinby is due West of Bronwich and is 123 km away.
 The bearing of Dinby from Carr is 307°.
 a) Draw a sketch to show this information.
 b) Calculate the distance between Bronwich and Carr. Give your answer to 3 significant figures.
 c) How far north of Bronwich is Annerly? Give your answer to 4 significant figures.
 d) Work out the distance and bearing of Annerly from Dinby.

CHAPTER 18

2-D and 3-D objects

 EXERCISE 18.1

1 A cuboid measures 5 cm by 6 cm by 2 cm.
  a) Draw a sketch of the cuboid.
  b) Make an isometric drawing of the cuboid.
  c) Draw a net for the cuboid.

2 A pyramid is drawn on a rectangular base.
  The base is 4 cm by 5 cm. The pyramid is 8 cm tall.
  a) Draw a sketch of the pyramid.
  b) Draw a plan view, a front elevation and a side elevation.
  c) Draw an accurate net for the pyramid.

3 A triangular prism is 8 cm long. Its ends are equilateral triangles of side 5 cm.
  a) Draw a sketch of the prism.   **b)** Draw an accurate net for the prism.

4 The isometric drawing shows some cubes forming a 3-D shape.
  a) Copy the shape onto isometric paper.
  b) On squared paper draw a sketch of
     (i) a plan view as seen from **A**
     (ii) a side elevation as seen from **B**
     (iii) a front elevation, as seen from **C**.

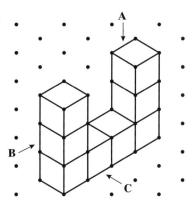

5 Kamini and Dakshesh have been building shapes with centimetre cubes on a square grid.
  The diagrams show plan views of their shapes.
  The numbers 1, 2, 3 tell you how many cubes are stacked on top of each square.

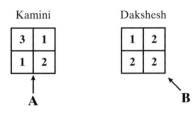

  a) Draw a front elevation to show how Kamini's shape appears seen from direction A.
  b) Make an isometric drawing of Dakshesh's shape, viewed from direction B.

6 Here are three projections of a solid object.

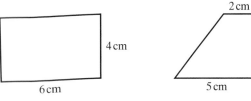

| Side elevation | Front elevation | Plan view |

a) Make a sketch of the solid object.
b) Make an isometric drawing of the solid object.

EXERCISE 18.2

1 The diagram shows a cuboid ABCDOEFG.
E has coordinates (2, 0, 0). B has coordinates (2, 4, 3).
a) Write down the coordinates of the points
A, C, D, F and G.
b) The point M is midway between O and B.
Find the coordinates of M.
c) The point N is midway between B and D.
Find the coordinates of N.

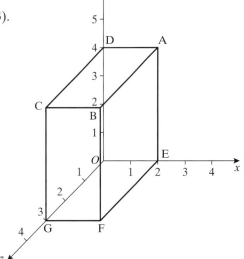

2 The diagram shows a solid OABC.
PA and QC are parallel to the x axis,
BC is parallel to the y axis, BA is parallel to
the z axis.
OB = 7 units, OQ = 6 units, OP = 3 units.
a) Write down the coordinates of A, B and C.

M is the midpoint of BC,
and N is the midpoint of AC.
b) Find the coordinates of M and N.
c) Use your answers to part **b)** to explain how
you can tell the line segment MN is horizontal.
d) What name best describes the solid OABC?

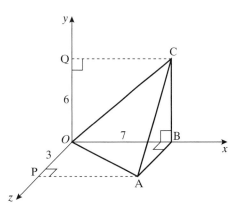

3 The diagram shows a cuboid ABCDEFGO.
   E is $(5, 0, 0)$, D is $(0, 3, 0)$, G is $(0, 0, 2)$.
   a) Write down the coordinates of B and C.

   M is the midpoint of BD.
   b) Find the coordinates of M.

   N is the midpoint of EG.
   c) Find the coordinates of N.
   d) **(i)** Write down the coordinates of the
           midpoint of MN.
       (ii) Explain how you could get these coordinates
               without working out the coordinates of M
               and N.

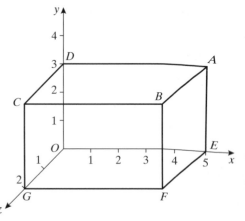

4 Point A is at $(2, 3, 0)$, B is at $(4, 11, 6)$ and C is at $(4, 9, 6)$.
   M is the midpoint of line segment AB.
   a) Work out the coordinates of the point M.

   N is the midpoint of line segment MC.
   b) Work out the coordinates of the point N.

5 Here are the coordinates of six points:
   A $(1, 3, 0)$,   B $(5, 2, 0)$,   C $(1, 5, 0)$,   D $(3, 4, 0)$,   E $(4, 3, 0)$,   F $(2, 0, 3)$.
   a) Which of the points is the midpoint of the line segment BD?
   b) What can you say about the triangle ACD?

EXERCISE 18.3

Find the volume and total surface area of each of the solids in questions **1** to **6**.
Give your answers exactly, in terms of π.
Also give corresponding numerical values, correct to 3 significant figures.

1
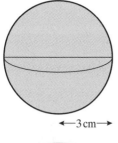
←3 cm→

2
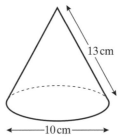
13 cm
←10 cm→

3
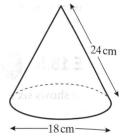
24 cm
←18 cm→

4
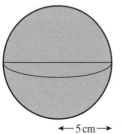
←5 cm→

5
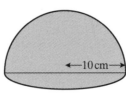
←10 cm→

6
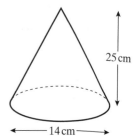
25 cm
←14 cm→

7   A pyramid has a square base of side 4 cm and a vertical height of 7 cm. Find its volume.

8   A square-based pyramid has a volume of 5000 cm³. Its height is 6 cm.
Find the dimensions of the base.

9   A sphere has surface area 484π cm². Find its volume in terms of π.

10 A cone of slant height 10 cm has a curved surface area of 50π cm².
Find the area of its base in terms of π.

EXERCISE 18.4

1   Convert 3.62 m³ into cm³.   **2**   Convert 280 mm² into cm².   **3**   Convert 200 000 m² into km².

4   Convert 3.85 cm³ into mm³.   **5**   Convert 490 000 mm² into m².

6   A pyramid has a volume of 2 600 000 mm³.   **7**   A cuboid has a surface area of 0.6 m².
Express its volume in cm³.   Express its area in cm².

For questions **8** to **10**, you may use the information that 1 litre = 1000 cm³.

8   A bath contains 450 litres of water.
    a)   Convert 450 litres into cm³.
    b)   Hence find the amount of water in the bath in m³.

9   A water tank in the shape of a cuboid measures 2.3 m by 5.7 m by 3 m.
    a)   Find the volume of the tank, in m³.
    b)   Convert this answer into cm³.
    c)   How many litres of water can the tank hold?

10 A tank in the shape of a cube has a capacity of 1331 litres.
    a)   Express this capacity in cm³.
    b)   Convert your answer to part **a)** into m³.
    c)   Find the dimensions of the tank in metres.

EXERCISE 18.5

1   The table shows six expressions: p, q and r are lengths' 5 and 2 are numbers and have no
dimension.

$p + 2q + r$	p^2q^2	$2rq^2$	$\dfrac{pr^2}{q}$	$5pqr$	$2p(r + 5q)$

    a)   Copy the table and put the letter A in the box underneath each of the **two** expressions that
    could represent an **area**.
    b)   Put the letter V in the box underneath each of the **two** expressions that could represent a
    volume.

2 Here are three expressions.

Expression	Length	Area	Volume	None of these
$3\pi x + \pi y^2$				
$\pi z^2 + 2y^2$				
$\pi x^2 y$				

x, y and z are lengths. π and 3 are numbers and have no dimensions.

Copy the grid and put a tick (✓) in the correct column to show whether the expression can be used for length, area, volume or none of these.

3 The expressions below can be used to calculate lengths, areas or volumes of some shapes.
The letters b, w and h represent lengths. π and 3 are numbers and have no dimension.
Pick out the **three** expressions that can be used to calculate a **volume**.

$bh + 3w$ $\qquad$ $\dfrac{\pi}{3}bwh$ $\qquad$ $\dfrac{bh}{3}$ $\qquad$ $\dfrac{w^2h^2}{b}$ $\qquad$ $\pi b^2 h^2$

$\pi h(b^2 + 3wh)$ $\qquad$ πbh $\qquad$ $3\pi(bw + wh)$ $\qquad$ $bh(\pi + 3)$ $\qquad$ $3b^2 + 3bh$

4 Here are three expressions.

Expression	Length	Area	Volume	None of these
$6\pi b(2a + c)$				
$\dfrac{3ac}{\pi b}$				
$\pi a^2 + 2\pi b$				

a, b and c are lengths. π, 2, 3 and 6 are numbers and have no dimensions.
Copy the grid and put a tick (✓) in the correct column to show whether the expression can be used for length, area, volume or none of these.

5 Here are some expressions. The letters x, y, z and h represent lengths.
π and 10 are numbers that have no dimensions.
Three of the expressions could represent areas.
Copy the grid and tick (✓) the boxes underneath the three expressions which could represent areas.

$\pi(x + y + z)$	$10\pi xh$	$\pi x(10h + z)$	y^2z	$\dfrac{\pi x^2 h}{y}$	$10y^3z$	$\dfrac{\pi yz}{x}$

CHAPTER 19

Circle theorems

EXERCISE 19.1

1 PT is a tangent to the circle, centre O. Angle POT = 67°.
  a) State, with a reason, the value of the angle marked *x*.
  b) Work out the value of the angle marked *y*.

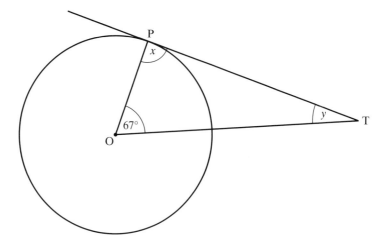

2 PT is a tangent to the circle, centre O.
PT = 12 cm. OP = 5 cm.
The line OT intersects the circle at R, as shown.
Work out the length of RT.

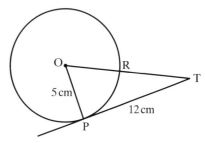

3 TP and TR are tangents to the circle, centre O. Angle POR is 148°.

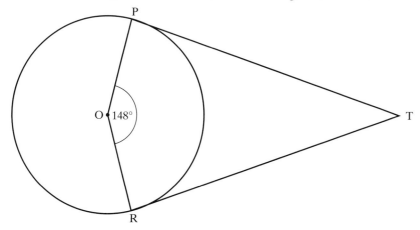

  a) Work out the size of angle PTR. Give reasons.
  b) What type of quadrilateral is OPTR? Explain your reasoning.

19 Circle theorems 79

4 TP and TR are tangents to the circle, centre O. Angle ORP is 10°.

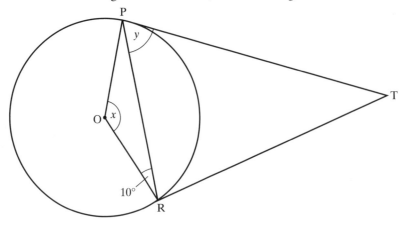

    a) Work out the value of x.
    b) Work out the value of y.
    c) What type of triangle is triangle RPT?

5 The diagram shows a circle, centre O.
   The chord AB is 24 cm.
   M is the midpoint of AB.
   OM = 7 cm.

   Calculate the length of the radius of the circle.

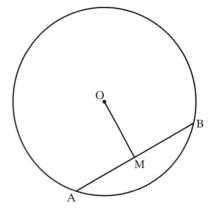

6 The diagram shows a circle, centre O, radius 30 cm.
   PQ = 48 cm.
   OM is perpendicular to PQ.

   Work out the length of OM.

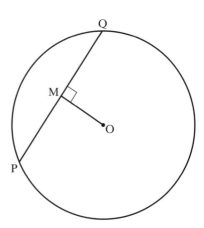

7 From a point A, two tangents AB and AC are drawn to a circle, centre O.
  a) Make a sketch to show this information.
  The angle BOC is measured, and found to be exactly 90°.
  b) What type of quadrilateral is OBAC?

8 The diagram shows a circle, centre O. AB and CD are chords.
  The diameter PQ passes through the midpoints M and N
  of the chords.
  OM = 7 cm, AB = 10 cm and CD = 13 cm.

  a) Explain why angle OMB = 90°.
  b) Use Pythagoras' theorem to calculate the
  distance OB. Show your working.
  c) Calculate the distance MN.

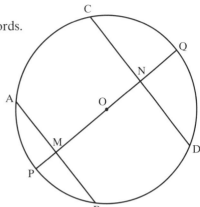

EXERCISE 19.2

Find the missing angles A to Z in these diagrams, which are not drawn to scale.
O is the centre of each circle.
Explain your reasoning in each case.

1

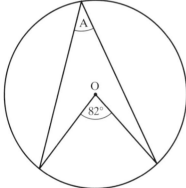

2

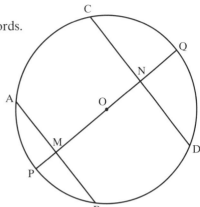

3

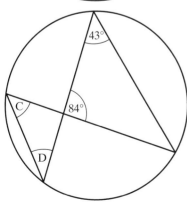

4

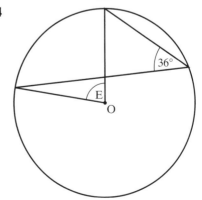

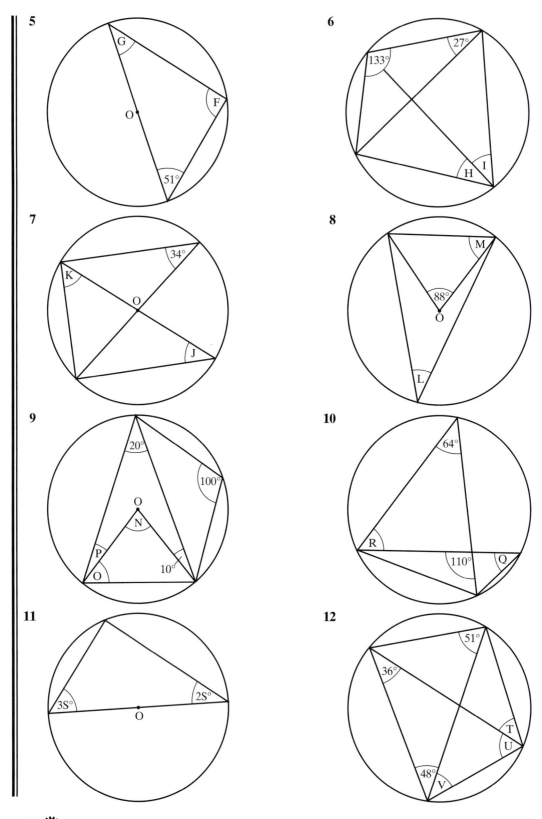

19 Circle theorems

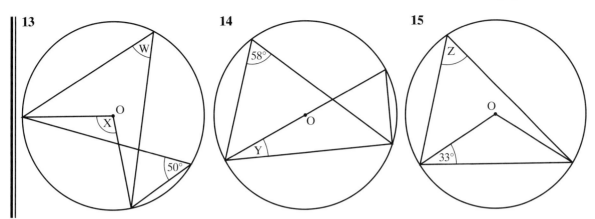

13

14

15

EXERCISE 19.3

Find the missing angles A to Z in these diagrams, which are not drawn to scale.

O is the centre of the circle.

Explain your reasoning in each case.

1

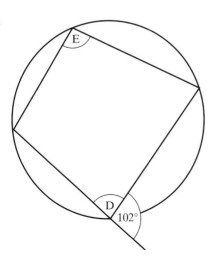

2

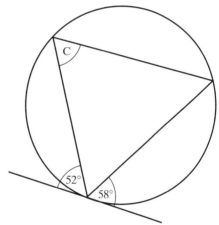

3

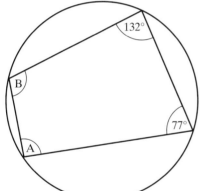

4

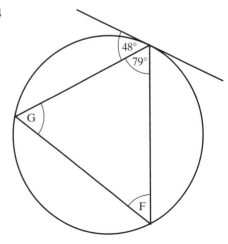

5

6

7

8

9

10

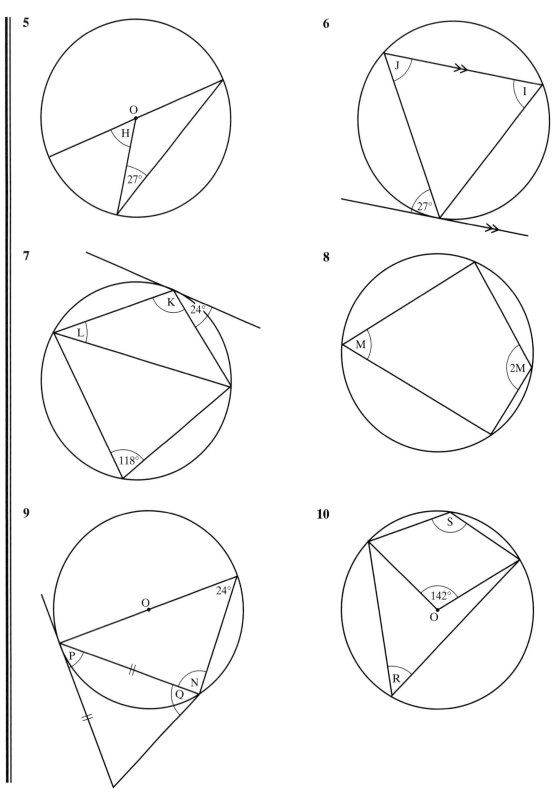

19 Circle theorems

11

12

13

14

15

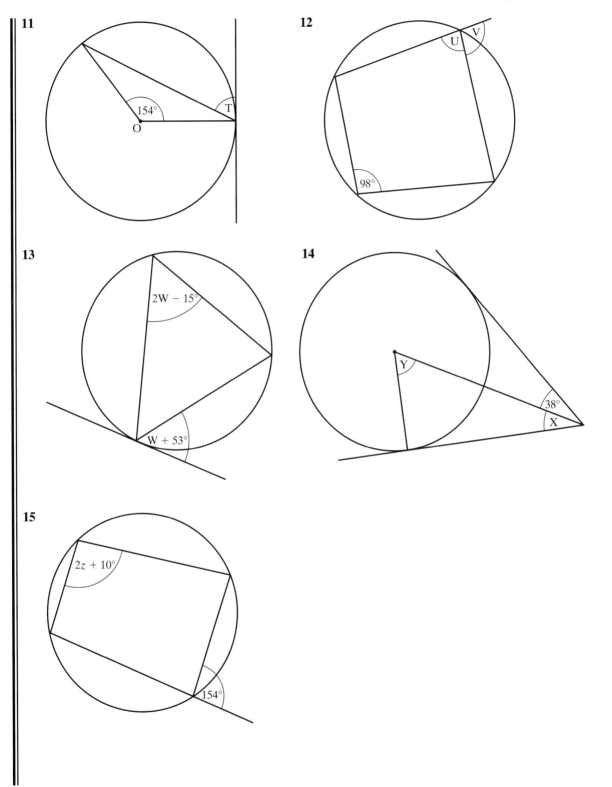

CHAPTER 20

Collecting data

 EXERCISE 20.1

1 Fran carries out a survey.
  She wants to see whether the school canteen should be open every day.
  She reaches the canteen at 12 noon and asks the first 50 students who arrive the question,
  'Don't you agree that our school canteen should stay open every day?'
  a) Write down **two** reasons why this is **not** a good way to find out whether students at Fran's
    school think the school canteen should be open every day.
  b) Describe a better method of selecting 50 students that Fran might use.
  c) Write down a replacement question that Fran could use.

2 A newspaper editor wants to find out people's opinions about a law that has just been passed in
  government.

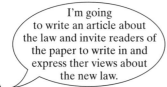

  I'm going to write an article about the law and invite readers of the paper to write in and express ther views about the new law.

  Reporter A

  Im going to stand in the street and ask every tenth person what they think of the new law.

  Reporter B

  Explain which of these methods is the better one.

3 Davin is studying the number of matches in boxes of Rover matches.
  He wants to see if the number of matches in each box is the same.
  He goes to his local newsagent and buys 50 boxes of Rover matches.
  He then takes them home and counts the numbers in each box.
  Explain briefly whether you think this is a good sampling method.

4 The table shows the number of students in each year group at Wendower School.
  There are 800 students at the school altogether.

Age group	Year 7	Year 8	Year 9	Year 10	Year 11
Number of students	185	199	226	82	108

  Mavis wants to take a stratified sample of 60 students from the school.
  Draw up a table to show how many students she should select from each year group.

5 David and Graham are planning a coursework exercise.
They want to find out about the student absences at different schools in London.

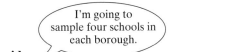

I'm going to sample four schools in each borough.

David

I'm going to sample 30% of the schools in London.

Graham

Explain why one of these methods is much better than the other.

6 Explain briefly the difference between *primary* and *secondary* data.
Here are some questions suitable for statistical investigation.
For each one, say whether you would most likely use primary or secondary data.
a) In your street, do people use the local library?
b) Does a student's final GCSE results relate to their SATs results at Key Stage 3?
c) Do people who work in hospitals tend to take more time off work than teachers?
d) Are people's weights totally related to their heights?
e) Are the children in your family older than the children in your friends' families?
f) Do the students in your class spend more time watching television than students in another class at your school?

EXERCISE 20.2

1 Fred is carrying out a survey.
He wants to use a questionnaire to find out what pets students at his school like.
He also wants to see if boys like the same pets as girls.
Write down two questions that Fred might use in his questionnaire.

2 Mel is carrying out a survey to see whether teachers and students like the same music.
She asks each person in the sample to name their favourite type of music.
Mel records the results on a data collection sheet.
Here are her results after asking the first fourteen people.

Pop	Classical	Jazz	Folk	Teachers	Students

a) Explain one disadvantage with recording the data in this way.
b) Design an improved data collection sheet that Mel might use.

3 Keith is conducting a survey about restaurants in his town.
Here are three questions in his questionnaire.

1	How old are you?	_____
2	What restaurant do you go to?	_____
3	Do you agree that restaurants are too expensive?	YES NO (circle one)

a) Explain briefly why each of Keith's questions is unsuitable.
b) Design suitable replacements for each of Keith's questions.

4 Jonathan has been carrying out a survey.
He asked some tea and coffee drinkers whether they drank tea or coffee or both.
Here are some of his results, transferred into a two-way table.

	Tea	Coffee	Both	Total
Male	15			42
Female		12	17	
Total			32	100

a) Complete the two-way table.
c) How many males were in the survey?

b) How many females drank only tea?
d) How many people only drank coffee?

5 Jean has collected data on how much weight (w), in kg, people have lost at her slimming club over the last 6 months.
This two-way table shows some of her results.

	Men	Women	Totals
$0 \leqslant w < 5$	1		10
$5 \leqslant w < 10$		26	
$10 \leqslant w < 15$		17	27
over 15	9	11	
Totals	27		

a) Complete the two-way table.
c) How many men lost 10 or more kg?
e) How many people took part in Jean's survey?

b) How many women lost over 15 kg?
d) How many women lost less than 10 kg?

6 60 students were asked if they liked maths.
35 of the students were boys.
48 students liked maths.
4 boys did *not* like maths.
Use this information to complete a copy of the two-way table.

	Liked maths	Did not like maths	Total
Boys			
Girls			
Total			

7 Design a data collection sheet for a survey about how much homework students get.

20 Collecting data

CHAPTER 21

Working with data

EXERCISE 21.1

1 A company sells beds in flat packs for customers to build.
This stem and leaf diagram shows the times, in minutes, it took for each of 14 people to build a bed.

```
2 | 4  5
3 | 3  8  8
4 | 0  1  7  9
5 | 2  5  6
6 | 1  7
```

Four more people build a bed from a flat pack.
Their times, in minutes are 52, 48, 69 and 52.
a) Draw a new stem and leaf diagram to include the four new people.
b) Add a key to the diagram.

2 Patsy has made an unordered stem and leaf diagram to show the weights of 23 loaves of bread in her bakery.
The times are in minutes.

```
20 | 6  4  1
21 | 7  4  0  6  7
22 | 3  9  5  4
23 | 9  8  5  8  3  8
24 | 6  6  4  9  0
```

a) Redraw the diagram so that it is fully ordered.
b) Add a key to the new diagram.

3 There were 15 pairs of trainers in a sale.
Their prices, in £, are shown below:

```
23    38    29    53    46    40    23    37
20    34    56    42    33    39    21
```

a) Draw a stem and leaf diagram to show this information.
Remember to include a key.
b) Work out the range of the prices.
c) Find the median price.

4 A cycling group went on a 13-day holiday.
Here are the number of kilometres they planned to cycle each day.

18	32	23	28	19	40	41
25	34	38	19	22	27	

a) Draw a stem and leaf diagram to show this information.
b) Work out the range of the data.
c) Find the median distance.

5 16 students sat an English examination. Here are their marks:

72	68	53	60	72	50	83	57
50	75	81	76	65	69	60	71

a) Draw a stem and leaf diagram to show this information.
b) Find the median of the marks.

 EXERCISE 21.2

1 Terri conducts a survey.
She finds out how many hours of TV students
watched last Monday.
The frequency table shows her results.

a) How many students took part
in Terri's survey?
b) Work out the mean number
of hours of TV watched.

Number of hours TV	Frequency
0	2
1	7
2	9
3	4
4	3
5 or more	0
Total	

2 Mario records the number of people at the gym each day in April.
The frequency table shows his results.

Number of people at the gym	Frequency	Midpoint
28 to 34	4	
35 to 41	8	
42 to 48	12	
49 to 55	6	

a) Calculate an estimate of the mean number of people at the gym each day in April.
b) State the modal class.
c) Find the class interval which contains the median.
d) Carlo says, 'The range is 30.'
Explain why Carlo must be **wrong**.

3 Hajra measures the weights of some babies and toddlers at a creche.
The frequency table shows her results.

Weight (w) in kg	Frequency (f)
$0 < w \leqslant 4$	5
$4 < w \leqslant 8$	6
$8 < w \leqslant 12$	7
$12 < w \leqslant 16$	2
$16 < w \leqslant 20$	1

Without using a calculator, work out an estimate for the mean weight of these babies and toddlers.
Show all your working clearly.

4 The frequency table shows the ages of the passengers on a cruise liner on the Atlantic.

Age (A) in years	Frequency
$0 < A \leqslant 20$	38
$20 < A \leqslant 40$	70
$40 < A \leqslant 60$	75
$60 < A \leqslant 80$	15
$80 < A \leqslant 100$	2

Work out an estimate of the mean age of the passengers on the cruise liner.
Give your answer correct to 3 significant figures.

5 Seamus solves some Sudoku puzzles in a book.
The frequency table shows the times he took to solve each puzzle.

Time (t) in minutes	Frequency
$5 \leqslant t < 10$	4
$10 \leqslant t < 15$	5
$15 \leqslant t < 20$	7
$20 \leqslant t < 25$	9
$25 \leqslant t < 30$	5

a) Calculate an estimate of the mean time Seamus took to solve the puzzles.
b) Explain briefly why your answer can only be an estimate.

EXERCISE 21.3

1 A football club carried out a survey of the ages of its 274 supporters at a football match.
The results of the survey are shown in the table.

Age group, x years	Frequency
$0 \leqslant x < 10$	24
$10 \leqslant x < 20$	36
$20 \leqslant x < 30$	60
$30 \leqslant x < 40$	52
$40 \leqslant x < 50$	30
$50 \leqslant x < 60$	44
$60 \leqslant x < 70$	28

a) Draw a frequency polygon to show this information.
Use graph paper with the *frequency* axis 12 cm high, numbered from 0 to 60, and the *age* axis 14 cm wide, numbered from 0 to 70.
b) Which is the modal class?
c) Find the class interval which contains the median.
d) Find an estimate for the mean age.

2 Justin is a florist.
He measured 100 flowers and recorded the heights to the nearest centimetre.
The table shows information about the heights, h, of the 100 flowers.

Height, h	Frequency, f
$0 \leqslant h < 20$	24
$20 \leqslant h < 40$	38
$40 \leqslant h < 60$	16
$60 \leqslant h < 80$	12
$80 \leqslant h < 100$	10

a) Draw a frequency polygon to show this information.
Use graph paper with the *frequency* axis 8 cm high, numbered from 0 to 40, and the *age* axis 10 cm wide, numbered from 0 to 100.
b) Work out an estimate for the mean height of these flowers.
c) Find the class interval that contains the median.

3 The table gives information about how long, in minutes, some people took to read a magazine. Use the table to draw a histogram that shows this information.

Time (t) in minutes	Frequency
$0 < t \leqslant 10$	20
$10 < t \leqslant 15$	30
$15 < t \leqslant 30$	60
$30 < t \leqslant 60$	90

4 A teacher asked some Year 10 students how long they spent doing homework each night. The histogram was drawn from this information.

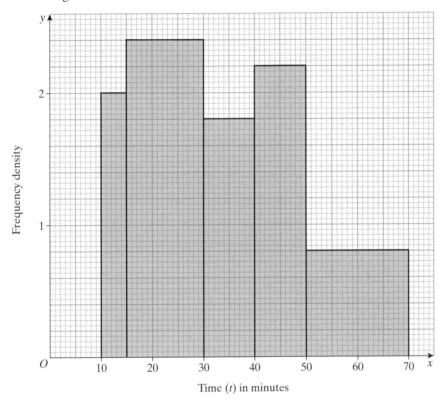

Time (t) in minutes

Use the histogram to complete the table.

Time (t) in minutes	Frequency
$10 \leqslant t < 15$	10
$15 \leqslant t < 30$	
$30 \leqslant t < 40$	
$40 \leqslant t < 50$	
$50 \leqslant t < 70$	

[Edexcel]

5 A teacher asked some students how much time they spent using a mobile phone one week. The histogram was drawn from this information.

Use the histogram to complete a copy of the table.

Time (t) in hours	Frequency
$0 \leqslant t < \frac{1}{2}$	
$\frac{1}{2} \leqslant t < 1$	
$1 \leqslant t < 2$	30
$2 \leqslant t < 3$	
$3 \leqslant t < 5$	

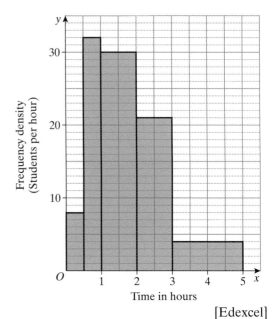

[Edexcel]

6 Some students at Highfliers School took a mathematics examination.
The unfinished table and histogram show some information about their marks.

Mark ($x\%$)	Frequency
$0 < x \leqslant 40$	10
$40 < x \leqslant 60$	40
$60 < x \leqslant 75$	45
$75 < x \leqslant 85$	60
$85 < x \leqslant 95$	
$95 < x \leqslant 100$	25

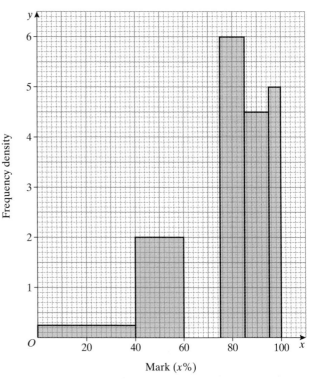

a) Use the information in the table to complete a copy of the histogram.
b) Use the information in the histogram to complete the table. [Edexcel]

EXERCISE 21.4

1 320 students took a test.
The cumulative frequency graph gives information about their marks.

Work out an estimate for the interquartile range of their marks.

[Edexcel]

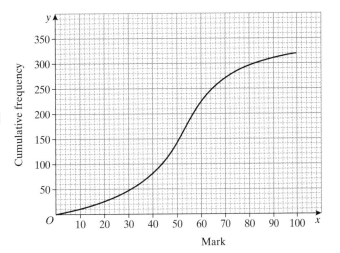

2 Bill asked 80 students how many hours each of them spent doing homework last week.
He used the information to draw this cumulative frequency graph.

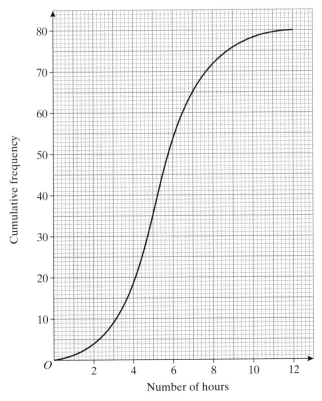

Use the cumulative frequency graph to find an estimate for:
a) the median number of hours
b) the number of students who spent more than 7 hours on their homework. [Edexcel]

3 The cumulative frequency diagram below gives information about the prices of 120 houses.
   a) Find an estimate for the number of houses with prices less than £130 000.
   b) Work out an estimate for the interquartile range of the prices of the 120 houses.

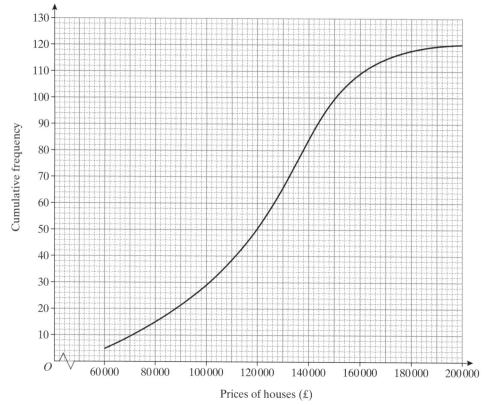

Prices of houses (£)

[Edexcel]

4 The table shows information about the number of hours that 120 children used a computer last week.

Number of hours (h)	Frequency
$0 < h \leqslant 2$	10
$2 < h \leqslant 4$	15
$4 < h \leqslant 6$	30
$6 < h \leqslant 8$	35
$8 < h \leqslant 10$	25
$10 < h \leqslant 12$	5

a) Complete a copy of the cumulative frequency table.

Number of hours (h)	Cumulative frequency
$0 < h \leqslant 2$	10
$2 < h \leqslant 4$	
$4 < h \leqslant 6$	
$6 < h \leqslant 8$	
$8 < h \leqslant 10$	
$10 < h \leqslant 12$	

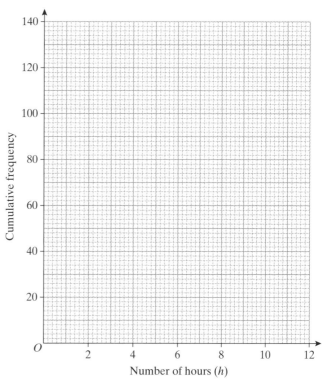

b) On a copy of the grid, draw a cumulative frequency graph for your table.

c) Use your graph to find an estimate for the number of children who used a computer for **less** than 7 hours last week.   [Edexcel]

5 The cumulative frequency table gives information about the heights, in centimetres, of some students at Bax School.
a) On a grid, with the horizontal axis having heights from 135 cm to 175 cm and the vertical axis having a cumulative frequency of 0 to 180, draw a cumulative frequency graph for the data in the table.
b) Use your cumulative frequency graph to estimate:
   (i) the median height
   (ii) the number of students with height less than 151 cm.

Heights (h) of students in cm	Cumulative frequency
$h < 135$	0
$135 \leqslant h < 140$	2
$135 \leqslant h < 145$	10
$135 \leqslant h < 150$	29
$135 \leqslant h < 155$	70
$135 \leqslant h < 160$	115
$135 \leqslant h < 165$	143
$135 \leqslant h < 170$	156
$135 < h \leqslant 175$	163

EXERCISE 21.5

1 Match these graphs with their descriptions:

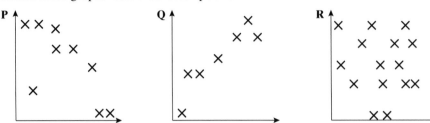

  P   Q   R

    A   Length of hair against Height
    B   Value of car against Age of car
    C   Height against Weight

2 A park has an outdoor swimming pool.
The scatter graph shows the maximum temperature and the number of people who used the pool on ten Saturdays in summer.

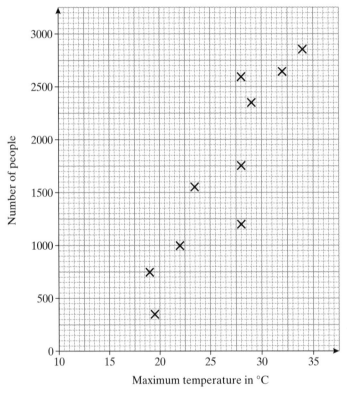

    a) Describe the correlation between the maximum temperature and the number of people who used the pool.
    b) Draw a line of best fit on the scatter graph.
    The weather forecast for the next Saturday gives a maximum temperature of 27 °C.
    c) Use your line of best fit to estimate the number of people who will use the pool.   [Edexcel]

3 Some students took a mathematics test and a science test.
The scatter graph shows information about the test marks of eight students.

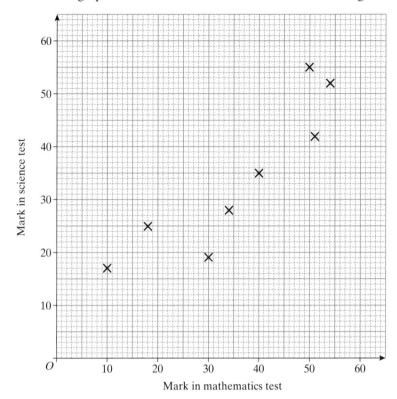

The table shows the test marks of four more students.

Mark in mathematics test	14	25	50	58
Mark in science test	21	23	38	51

a) On a copy of the scatter graph, plot the information from the table.
b) Draw a line of best fit on your scatter graph.
c) Describe the **correlation** between the marks in the mathematics test and the marks in the science test.

[Edexcel]

4 Ten men took part in a long jump competition.
The table shows the heights of the ten men and the best jumps they made.

Best jump (m)	5.33	6.00	5.00	5.95	4.80	5.72	4.60	5.80	4.40	5.04
Height (m)	1.70	1.80	1.65	1.75	1.65	1.74	1.60	1.75	1.60	1.67

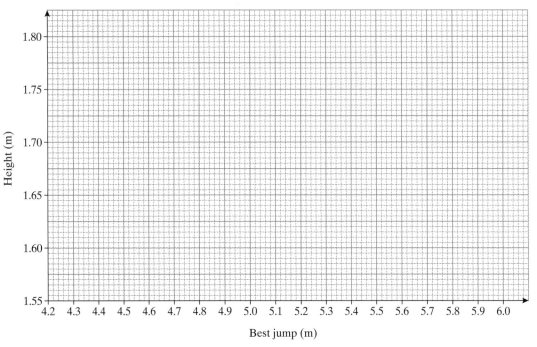

a) On a copy of the grid above, plot the points as a scatter diagram.
b) Describe the relationship between the height and the maximum distance.
c) Draw in a line of best fit.   [Edexcel]

5 Here is a scatter graph.
One axis is labelled 'Height'.
a) For this graph, state the type of correlation.

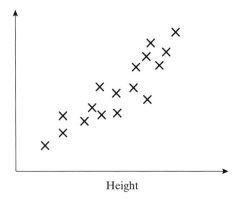

b) From the list below, choose the most appropriate label for the other axis:

Length of hair   Number of sisters   Length of legs   GCSE French mark

[Edexcel]

CHAPTER 22

Probability

EXERCISE 22.1

1 A bag contains 27 balls.
There are 10 blue balls and 13 red balls.
The rest of the balls are black.
A ball is chosen at random.
Find the probability that this ball is

   a) red   **b)** **not** blue   **c)** black.

2 The colours on a regular 5-sided spinner are red, blue, yellow, green and purple.
Dean spins the spinner once.
Find the probability that it

   a) lands on red   **b)** does **not** land on green   **c)** lands on black.

3 The probability that I will be at school tomorrow is 0.93.
Work out the probability that I will **not** be at school tomorrow.

4 There are 500 sweets in a jar.
100 of the sweets are red.
The rest are orange, yellow or green.
Ted takes a sweet at random from the jar and puts it back.
   a) Calculate the probability that Ted takes a red sweet.
Tessa takes 50 sweets at random from the jar.
20 of these sweets are orange.
   b) Estimate how many orange sweets are in the jar.

5 A drawer contains 40 socks.
10 of them are black, the rest are white or grey.
There are four times as many white socks as grey socks.
A sock is chosen at random.
   a) Work out the probability that it is black.
   b) Work out the probability that it is **not** black.
   c) Work out the probability that it is white.

6 The two-way table shows some information about the number of people in a library.

	Adults	Children	Total
Male		7	18
Female	20		
Total			50

a) Copy and complete the two-way table.
b) A person in the library chooses a book.
  Work out the probability that the person is a child.
c) One of the people in the library uses the computer.
  Work out the probability that the person is male.

7 A box contains only 1p coins, 2p coins and 5p coins.
  There are 200 coins in the box and 50 of these are 2p coins.
  Miriam randomly takes a coin from the box and then puts it back.
  a) Work out the probability that the coin Miriam picked was a 2p coin.
  Miriam then randomly takes 40 coins from the box. 10 of these are 5p coins.
  b) Estimate the total number of 5p coins in the box.

8 A small playgroup has children aged 2 years, 3 years and 4 years.
  Darren wrote down the ages of the children that attended on Tuesday.
  The two-way table shows the results.

	2 years	3 years	4 years	Total
Boys	5	9	5	19
Girls	6	5	7	18
Total	11	14	12	37

  a) Find the probability that a child is a 2-year-old girl.
  b) Find the probability that a 4-year-old is a boy.
  c) Find the probability that a child is 3 years old.

9 A restaurant serves fruit salad, cheesecake or mousse for dessert.
  Mr Chen keeps a record of what he serves on Sunday.
  Every customer had one dessert.
  The table shows some of this information.

	Fruit Salad	Cheesecake	Mousse	Total
Midday	10			
Evening		8		46
Total	28		25	70

  a) Copy and complete the two-way table.
  b) Find the probability that a customer had cheesecake for dessert.
  c) Find the probability that a customer had mousse for dessert in the evening.

10 In the sixth form at Gravitas School there are 63 boys and 78 girls.
  a) A student is chosen at random from the sixth form.
    Work out the probability that the student is a girl.
  b) There are 1200 students altogether at Gravitas School.
    Estimate the total number of girls at the school.
  c) Explain why your estimate might not be very reliable.

EXERCISE 22.2

1 A biased dice shows scores of 1, 2, 3, 4, 5, 6 with these probabilities.

Score	1	2	3	4	5	6
Probability	0.23		0.3	0.18	0.1	0.05

The dice is rolled once. Find the probability that the score obtained is
a) 2   **b)** not 4   **c)** 5 or 6   **d)** an odd number.

2 When Andrew goes to work he is either early, on time or late.
The incomplete table shows some probabilities.

	Early	On time	Late
Probability	$\frac{27}{100}$		$\frac{6}{100}$

a) What is the probability that Andrew arrives on time?
b) Which of the three things is Andrew most likely to do?
c) Find the probability that he is **not** late for work.

3 There are four types of animals I might see in my garden.
The table shows the probability that a fox or a hedgehog or a squirrel or a cat will be seen in my garden. No two animals are ever in the garden at the same time.

Type of creature	Fox	Hedgehog	Squirrel	Cat
Probability	0.15		0.4	0.28

a) Copy and complete the probability table.
b) Which type of animal is the most common in my garden?

An animal is observed at random. Find the probability that:
c) it is **not** a cat
d) it is a fox or a squirrel.

4 Bertha has three children, Robert, Johnny and Stephen.
Every Friday night Bertha has dinner with one of her children.
The probability that she has dinner with Robert is 0.28 and the probability that she has dinner with Johnny is 0.3.
a) Work out the probability that she has dinner with Stephen.
b) Work out the probability that she does **not** have dinner with Stephen.

5 A biased spinner gives scores of 1, 2, 3 or 4.
The probability of getting 1 is 0.37.
The probability of getting 2 is 0.18.
The probability of getting 4 is 0.2.
a) Calculate the probability of getting an even score.
b) Work out the probability of getting a score of 3.

6 Sue often partners Janet or Don in tennis.
The probability that she will partner Janet is 0.38.
The probability that she will partner Don is 0.17.
a) Work out the probability that she will partner Janet or Don in tennis.
b) Work out the probability that she will **not** partner either of these two in tennis.

7 Alex goes to the cinema.
He can choose from a comedy, a drama, a science fiction movie or a horror movie.
The table shows the probability that Alex will choose a particular type of movie.

Type of movie	Comedy	Drama	Science fiction	Horror
Probability	0.17	0.23	x	x

a) Find the probability that he chooses a comedy or a drama.
b) Find the probability that he chooses a science fiction movie or a horror movie.
The probability he will choose a science fiction movie is the same as the probability he will choose a horror movie.
c) Find the probability he will choose a horror movie.

8 The probability that Martin will be late for work on Monday is 0.8.
The probability that Lisa will be late for work on Monday is 0.4.

Give two reasons why Jamie is not correct.

The probability that they will both be late to work on Monday is 1.2

Jamie

9 The junior football team at Roddington School is picked from students in Years 7 or 8 or 9.
Mr Right selects the junior football teams. The probability that he might select a student from Years 7 or 8 is shown in the table below.

Year group	Year 7	Year 8	Year 9
Probability	22%	32%	

a) Copy and complete the probability table.
b) Mr Right needs to pick 50 students for some junior football teams.
Estimate the number of Year 8 students he is likely to pick.

10 A drawer contains black, white, blue and brown socks.
In a probability experiment, one sock is chosen at random, and removed from the drawer.
Its colour is noted, and it is returned to the drawer.
The table shows some probabilities for this experiment.

Colour	Black	White	Blue	Brown
Probability	0.36	0.32	0.2	

a) Find the probability of obtaining a brown sock.

The experiment is carried out 300 times.

b) Estimate the number of times a white sock is obtained.

c) Rewrite the probability table to show the probabilities as fractions with the same denominator.

d) Altogether the drawer contains x socks.
What is the smallest possible value of x?

 EXERCISE 22.3

1 The probability that Brad will be injured in a football match is 0.13.
The probability that Marc will be injured in a football match is 0.4.
Calculate the probability that:
a) Brad and Marc will both be injured in a football match
b) Brad will be injured and Marc will not be injured in a football match.

2 A fair spinner has four sides numbered 2, 3, 4 and 5.
It is spun twice and the numbers are added together.
a) Draw up a two-way table to show the possible combinations of scores.
b) Find the probability that the total score is 8.
c) What is the most likely total score?

3 A bag contains 5 blue marbles, 2 red marbles and 6 green marbles.
A marble is chosen at random. It is then replaced, and a second marble is chosen.
a) Work out the probability that the first marble is **not** red.
b) Work out the probability that both marbles are blue.

4 Terri has a three-sided spinner, with the numbers 3, 4 and 5 written on it.
She also has a four-sided spinner, with the numbers 2, 3, 5 and 6 written on it.
Terri spins both spinners and multiplies the two numbers together.
a) Draw up a two-way table to show the possible results.
b) Work out the probability that the product of the two numbers is 15.
c) Work out the probability that the product of the two numbers is even.

5 Garth throws a big dice and a small dice.
Both dice are normal fair dice, labelled with the numbers 1, 2, 3, 4, 5, 6.
a) Write down the probability that the small dice shows a number bigger than 6.
b) Work out the probability that the big dice shows a larger number than the small dice.
c) Work out the probability that Garth throws two even numbers.

6 A box contains coloured pencils which are red or blue or pink.
Jilly takes a pencil from the box at random and puts it back.
The probability that the pencil is red is 0.6.
The probability that it is pink is 0.1.
a) Write down the probability that Jilly takes a blue pencil.

Ian takes two pencils from the box and puts them back.
b) Work out the probability that Ian takes two red pencils.

Benji takes three pencils from the box.
c) Work out the probability that the first two pencils are the same colour but the third one is not.

7 In a large box of light bulbs, $\frac{1}{20}$ are defective.
The manager of the store takes two light bulbs from the box.
  a) Calculate the probability that the first light bulb is defective and the second one is not.
  b) Calculate the probability that they are both defective.
Give your answers as exact fractions.

8 Each school day Mrs Kothare selects one girl and one boy at random, from some classes, to show her their homework diaries.
The table shows the probabilities of each student being chosen.

Girls	Year 7	Year 8	Year 9	Year 10
Probability	0.3	0.16	0.42	0.12

Boys	Year 7	Year 8	Year 9	Year 10
Probability	0.2	0.29	0.21	0.3

  a) Calculate the probability that Mrs Kothare chooses a girl from Year 9 and a boy from Year 10.
  b) Calculate the probability that she chooses a girl and boy from Year 10.
There are 200 school days one year.
  c) Estimate the number of days Mrs Kothare chooses two Year 7 students.

 EXERCISE 22.4

1 Patrick goes for two job interviews. The probability he will be offered the first job is 0.3 and the probability he will be offered the second job is 0.12.
  a) Copy and complete the tree diagram to show the possible outcomes of Patrick's job interviews.

First job   Second job

```
                    0.12 ──── Job offer
        0.3 ╱ Job offer ⟨
           ╱             ...  ──── No job offer
          ⟨
           ╲             ...  ──── Job offer
        ... ╲ No job offer ⟨
                    ...  ──── No job offer
```

  b) Work out the probability that Patrick is offered both jobs.
  c) Work out the probability that Patrick is offered only one job.

2 Sabia drives to work each morning.
She passes two sets of traffic lights on the way to work.
The probability that a set of traffic lights is green is $\frac{3}{5}$.
  a) Draw a tree diagram to represent the possible outcomes of the two traffic lights.
  b) Work out the probability that both sets of traffic lights are green.
  c) Work out the probability that the first set of traffic lights is **not** green and the second set of traffic lights is green.
  d) Work out the probability that just one of the two sets of traffic lights is green.

3 Jean sells two pairs of glasses.
The glasses have either single vision lenses, bifocal lenses or varifocal lenses.
The probability that Jean will sell glasses with single vision lenses is 0.4.
The probability that Jean will sell glasses with bifocal lenses is 0.35.

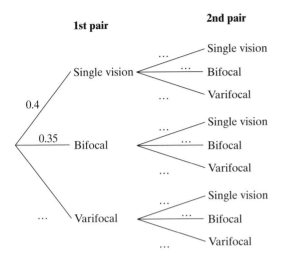

a) Copy and complete the tree diagram.
b) Find the probability that Jean sells one pair of glasses with bifocal lenses and one pair with varifocal lenses.
c) Find the probability that Jean does **not** sell two pairs of single vision lenses.

4 Jodi has 30 DVDs, 6 of which are musicals.
Wayne has 40 DVDs, 5 of which are musicals.
Jodi and Wayne choose one DVD each at random from their own collections.
a) Write down the probability that Jodi chooses a musical.
b) Write down the probability that Wayne chooses a musical.
c) Draw a tree diagram to illustrate their choices.
d) Hence find the probability that at least one of them does **not** choose a musical.

5 Louis is a fantastic golfer.
The probability that he gets a hole-in-one (the ball in the hole with one hit) is 0.02.
Louis plays three holes at golf.
a) Illustrate the possible results of getting a hole-in-one, or not, on a tree diagram.
b) Work out the probability that he gets three holes-in-one.
c) Work out the probability that he gets exactly one hole-in-one.

CHAPTER 23

Using a calculator efficiently

 EXERCISE 23.1

Use your calculator to work out the values of these quantities.
Give your answers correct to 3 significant figures where appropriate.

1 132^2	**2** 9^4	**3** 380^6	**4** 13^5
5 $\sqrt{20}$	**6** $\sqrt[3]{83}$	**7** $\sqrt[8]{3000}$	**8** $\sqrt[3]{71}$
9 23^{-2}	**10** 3.2^{-1}	**11** 0.5^{-1}	**12** $\frac{2}{7}$
13 $18^{\frac{1}{4}}$	**14** $228^{\frac{2}{3}}$	**15** $60^{-\frac{3}{4}}$	**16** $\sqrt[4]{75}$

17 Find the square of the square of 11.

18 Find the square root of the square root of 163.

19 Find the square root of the cube root of 85.

20 Find the reciprocal of the reciprocal of 27.3.

 EXERCISE 23.2

Use your calculator to work out the values of the following expressions. Write down all the figures on your calculator display, then round the answer to 3 significant figures where appropriate.

1 $(5.2 - 8.7 \times 0.3)^2 \div 6.2$ **2** $\sqrt{16.8 \div 2.7 + 18}$ **3** $(18.7 - 72.1 \div 5.8)^3$

4 $\dfrac{17.8 \times 0.032}{17.8 - 8.4}$ **5** $\dfrac{7.6 \div 1.9^3}{0.31 + 0.2}$ **6** $(3.87 + 2.1 \times 4)^{\frac{2}{3}}$

7 $\dfrac{53.5}{18.7 - 6.1 \times 3.9}$ **8** $\sqrt{\dfrac{3.7 \times 2.2}{6 + 0.3^2}}$ **9** $187 \div (9.03 + \sqrt{20.6})$

10 $17.1 \times (9.8 - 3.7)^2$ **11** $\sqrt[3]{3.8^2 - 1.4^2}$ **12** $\dfrac{7^3 + 2^2}{6^2 - 4^2}$

13 $\dfrac{22 - 1.82^2}{123 - 10.1^2}$ **14** $\sqrt{57.2^2 - 46.1^2}$ **15** $\dfrac{18.3 \times 1.1}{8.2 \times 2.8 - 3.1 \times 1.3}$

16 $\dfrac{6.2^2}{3.8^2 + 1.3^2}$

Use your calculator to work out the value of each of the following:

17 $\dfrac{8.7 \times 10^{15} + 9.6 \times 10^{12}}{3.4 \times 10^7}$

Give your answer in standard form correct to 3 significant figures.

18 $\sqrt[3]{9.15 \times 10^{28}}$

Give your answer in standard form correct to 2 significant figures.

19 $\dfrac{6.4 \times 10^{17}}{9.32 \times 10^{-3} - 8.15 \times 10^{15}}$

Give your answer in standard form correct to 3 significant figures.

20 $\sqrt{2.19 \times 10^{-4} + 7.83 \times 10^{15}}$

Give your answer in standard form correct to 3 significant figures.

EXERCISE 23.3

Use your calculator's fraction key to work out the answers to the following calculations.
Give your answers as ordinary fractions or mixed fractions as appropriate.

1 $\frac{3}{8} + \frac{4}{7}$ **2** $\frac{7}{9} \times \frac{5}{11}$ **3** $\frac{3}{4} - \frac{4}{17}$ **4** $\frac{7}{27} \div \frac{5}{6}$ **5** $6\frac{5}{8} + 8\frac{7}{11}$

6 $8\frac{2}{5} - 6\frac{5}{7}$ **7** $3\frac{2}{11} \times 2\frac{3}{5}$ **8** $2\frac{1}{7} \div 1\frac{1}{4}$ **9** $\left(3\frac{2}{9}\right)^2$ **10** $\sqrt{2\frac{7}{9}}$

Use your calculator to find the answers to these calculations as decimals.
Then convert the answers into exact fractions, using your fraction key.

11 6.4×2.6 **12** 0.75×13.2 **13** $51.875 \div 8.3$

14 $\sqrt{6.0025}$ **15** $15.08 - 2.8^2$ **16** $4.5^2 - 2.25$

17 $(13.9 + 17.6) \div 3.6$ **18** $(15^2 + 9^2 + 1^2) \times 0.02$ **19** $\sqrt[4]{133.6336}$

20 $14.85 \div 3.6$

EXERCISE 23.4

1 Maria invests £300 at 4% compound interest.
Work out how much her investment is worth:
 a) after 1 year **b)** after 2 years **c)** after 12 years.

2 Penny buys a new car for £20 000. At the end of each year, the value of the car has fallen to 70% of its value at the beginning of that year.
Work out how much Penny's car is worth:
 a) after 1 year **b)** after 2 years **c)** after 8 years.

3 Ariel invests £470 at 5% compound interest.
Work out how many years it takes until his investment has reached £800.

4 Tamsin pays £2400 for a new computer. At the end of each year, the value of the computer has fallen to 65% of its value at the beginning of that year. Tamsin decides to replace her computer once its value has fallen below £500.
Work out how many years it takes until Tamsin replaces her computer.

5 Kevin decides to put £320 into a savings scheme. The scheme pays 6% compound interest for the first year, then 8% compound interest per annum (each year) after that. Work out how much Kevin's savings are worth:
a) after 1 year   **b)** after 2 years   **c)** after 20 years.

6 Here are some instructions for making a number sequence:
- the first term is 4
- to make each new term, multiply the previous one by 2 and add 7.

a) Work out the first three terms of the number sequence.

b) Use your calculator's [ANS] key to help find the value of the 20th term.

 EXERCISE 23.5

1 A square has sides of length 9 cm, correct to the nearest centimetre.
a) Calculate the upper and lower bounds for the perimeter of the square.
b) Calculate the upper and lower bounds for the area of the square.

2 A rectangle has a length of 5.3 cm (rounded to 2 significant figures) and a width of 2.15 cm (rounded to 3 significant figures).
a) Calculate the upper bound for the perimeter of the rectangle.
b) Calculate the lower bound for the area of the rectangle.

3 To 1 decimal place, $x = 21.3$ and $y = 5.6$
a) Calculate the upper bound for the value of xy.
b) Calculate the lower bound for the value of $\dfrac{x}{y}$.
Write down all the numbers on your calculator display.

4 A cylinder has a radius of 5 cm and a height of 12 cm, both correct to the nearest centimetre. Calculate the maximum volume of the cylinder.

5 A = 12.2 (rounded to 1 decimal place)
B = 200 (rounded to the nearest hundred)
C = 8 (rounded to the nearest whole number)

a) Calculate the lower bound for A + B.
b) Calculate the upper bound for B ÷ C.
c) Calculate the maximum value of AC.
d) Calculate the least value of $\dfrac{A + B}{B + C}$.

CHAPTER 24

Direct and inverse proportion

EXERCISE 24.1

1 y is directly proportional to x, and $y = 8$ when $x = 10$. Find the value of y when $x = 15$.

2 y is directly proportional to x, and $y = 20$ when $x = 6$. Find x when $y = 50$.

3 Each of the tables below shows a set of matching x and y values, where y is directly proportional to x.
   Find a formula for y in terms of x, and work out the missing values in each case.

a)

x	2	8	10
y	6	24	

b)

x	1		5	20
y		20		100

c)

x	8	24	
y		6	11

d)

x		7		70	84
y		3	9		

4 y is directly proportional to x, and it is known that $y = 14$ when $x = 35$.
   a) Obtain an equation for y in terms of x.
   b) Use your equation to find the values of:
      (i) y, when $x = 42$ (ii) x, when $y = 60$.

5 y is directly proportional to the square root of x, and $y = 28$ when $x = 16$.
   Find y when $x = 400$.

6 y is directly proportional to x^2, and it is known that $y = 36$ when $x = 3$.
   a) Obtain an equation for y in terms of x.
   b) Use your equation to find the values of:
      (i) y, when $x = 10$ (ii) x, when $y = 100$.

7 y is directly proportional to x^3 and $y = 8$ when $x = 1$. Find y when $x = 6$.

8 Jason has a box of cylindrical shapes all made from the same material.
   The shapes are of different sizes.
   The weight, W grams, of each of these shapes varies as the cube of the length (x cm) of one of its sides.
   One of the shapes has a length of 3 cm and a weight of 108 grams.
   a) Find a formula for W in terms of x.
   b) Work out the weight of a shape with length 6.5 cm.
   c) One of the shapes weighs 2.916 kg. Work out the length of this shape.

9 Pierre sells jeans. His selling price, S, is directly proportional to the square of his cost price, C.
Pierre bought 30 pairs of jeans for £600 and sold each pair for £50.
 a) Find a formula that Pierre can use to work out his selling price from his cost price.
 b) Pierre buys 40 pairs of designer jeans for £1200.
   Use the formula to work out his selling price for each of these pairs of designer jeans.

10 The time, T seconds, of each swing of a simple pendulum varies directly as the square root of its length, L cm.
Chanelle has a 25 cm long pendulum that has a swing time of 1 second.
Robert has a pendulum that is four times as long as Chanelle's pendulum.
Calculate the swing time of Robert's pendulum.

EXERCISE 24.2

1 y is inversely proportional to x, and $y = 40$ when $x = 2$.
 Find the value of y when $x = 5$.

2 y is inversely proportional to x, and $y = 5$ when $x = 3$.
 Find x when $y = 3$.

3 Each of the tables below shows a set of matching x and y values, where y is inversely proportional to x.
 Find a formula for y in terms of x, and work out the missing values in each case.

a)

x	1		5
y	20	10	

b)

x	6	10	
y		24	16

c)

x	2		10
y		0.5	0.2

d)

x		5		20
y	5	2	1	

4 P is inversely proportional to Q, and $P = 12$ when $Q = 12$.
 a) Obtain an equation for P in terms of Q.
 b) Use your equation to find the values of:
   (i) Q, when $P = 36$
   (ii) P, when $Q = 9$.

5 R is inversely proportional to s, and it is known that $R = 9$ when $s = 40$.
 a) Obtain an equation for R in terms of s.
 b) Use your equation to find the values of:
   (i) R, when $s = 60$
   (ii) s, when $R = 18$.

6 y is inversely proportional to the cube of x, and $y = 25$ when $x = 2$. Find y when $x = 5$.

7 y is inversely proportional to x^2, and $y = 9$ when $x = 3$. Find x when $y = 81$.

8 The electrical resistance, R ohms, of a wire of given length is inversely proportional to the square of the diameter, d mm, of the wire.
When the diameter of the wire is 5 mm the resistance is 0.68 ohms.
   a) Find a formula for R in terms of d.
   b) Find the resistance of a wire with a diameter of 10 mm.

9 The volume, V cm³, of a gas is inversely proportional to its pressure, P cm of mercury.
When the volume is 100 cm³, the pressure is 54 cm of mercury.
   a) Calculate the volume when the pressure is 18 cm of mercury.
   b) Calculate the pressure when the volume is 270 cm³.

10 Fred has a car showroom. He decides that the selling price, £S, of each car in the showroom will be inversely proportional to the square root of the age, A years, of the car. He sells a 9-year-old car for £900.
   a) Work out the selling price of a 4-year-old car.
   b) What is the **difference** in selling price between the 1-year-old car and the 2-year- old car?
   Give your answer correct to the nearest £.

EXERCISE 24.3

Copy these six sketch-graphs into your book.
Alongside each, write either 'Direct proportion', 'Inverse proportion' or 'Neither'.

1
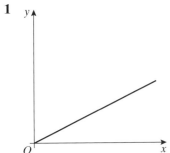

2
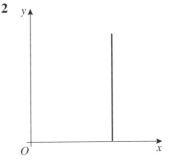

3
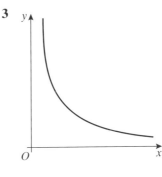

4
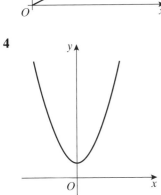

5
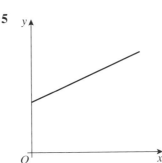

6
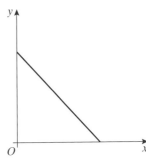

CHAPTER 25

Quadratic equations

 EXERCISE 25.1

Solve each of these quadratic equations by using the factorisation method.

1 $x^2 + 9x + 14 = 0$ **2** $x^2 + 19x + 60 = 0$ **3** $x^2 + x - 12 = 0$

4 $x^2 + 5x - 14 = 0$ **5** $x^2 + x - 30 = 0$ **6** $x^2 - 6x + 8 = 0$

7 $x^2 - 3x + 40 = 0$ **8** $x^2 - 12x + 35 = 0$ **9** $x^2 - 6x - 27 = 0$

10 $x^2 - x - 42 = 0$ **11** $2x^2 - 11x + 12 = 0$ **12** $3x^2 - 13x - 10 = 0$

13 $3x^2 + 8x + 4 = 0$ **14** $3x^2 + 2x - 16 = 0$ **15** $3x^2 + 13x + 4 = 0$

16 $2x^2 - 7x + 3 = 0$ **17** $2x^2 - 21x + 10 = 0$ **18** $5x^2 - 17x + 6 = 0$

19 $6x^2 + 17x + 12 = 0$ **20** $4x^2 + 7x + 3 = 0$

Here are some more difficult quadratic equations. Solve them by the factorisation method.

21 $4x^2 + 17x - 15 = 0$ **22** $6x^2 + 7x = 0$ **23** $25x^2 - 1 = 0$

24 $6x^2 + 9x = 0$ **25** $9x^2 - 12x + 4 = 0$ **26** $8x^2 - 2x = 0$

27 $6x^2 - 19x + 10 = 0$ **28** $20x^2 - 3x - 56 = 0$ **29** $9x^2 + 18x + 8 = 0$

30 $100 - 9x^2 = 0$

Rearrange these quadratic equations so that the right-hand side is zero. Then solve them by factorisation.

31 $2x^2 + x = 1$ **32** $x^2 + 24 = 10x$ **33** $x^2 + 19x = 11x + 20$

34 $x^2 + 7x = 2x + 6$ **35** $2x^2 = 7x - 3$ **36** $5 + 19x = 4x^2$

37 $6 = 7x + 5x^2$ **38** $3x^2 = 10x + 8$ **39** $6x^2 + 40x = 9x - 5$

40 $7x^2 = 44x - 12$

EXERCISE 25.2

Solve these equations using the quadratic equation formula.
Give your answers correct to 3 significant figures.

1 $x^2 + 4x + 2 = 0$ **2** $x^2 + 5x + 3 = 0$ **3** $2x^2 + 3x - 4 = 0$

4 $4x^2 - x - 4 = 0$ **5** $x^2 - 4x - 7 = 0$ **6** $2x^2 - 3x - 7 = 0$

7 $3x^2 - 8x + 2 = 0$ **8** $7x^2 + 8x - 2 = 0$ **9** $x^2 - 5x + 2 = 0$

10 $x^2 + 7x + 5 = 0$

Rearrange the equations below so that they are in the form $ax^2 + bx + c = 0$.
Then solve them using the formula method. Give your answers correct to 3 significant figures.

11 $5x^2 + 6x = 3$ **12** $x^2 + 2 = 4x$ **13** $2x^2 = 3x + 6$

14 $x(2x + 5) = 10$ **15** $2x(x - 1) = 5$ **16** $x^2 + x = 1$

17 $8x - 6 = x^2$ **18** $3x^2 = 20 + 2x$ **19** $2x^2 = 10 - 3x$

20 $4 + 2x = 7x^2$

EXERCISE 25.3

1 Two whole numbers y and $y + 3$ are multiplied together. The result is 108.
    a) Write down an equation in y.
    b) Show that this equation can be expressed as $y^2 + 3y - 108 = 0$.
    c) Solve the equation, to find the values of the two whole numbers.
    (There are two possible sets of answers, and you should give both.)

2 The diagram shows a triangle.
    Its base is $(4x - 6)$ cm and its height is $3x$ cm.
    Its area is 105 cm².
    a) Write down an equation in x.
    b) Show that this equation can be expressed as
        $2x^2 - 3x - 35 = 0$.
    c) x is a whole number. Solve the equation to find the
        value of x.
    d) Hence find the base and height of the triangle.

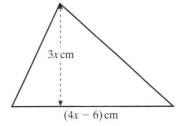

3 Chris and Hajra each thought of a positive whole number. Chris's number was 4 more than
    Hajra's number. Let Hajra's number be represented by x.
    a) Their two numbers multiply together to make 320. Write down an equation in x.
    b) Show that this equation can be expressed as $x^2 + 4x - 320 = 0$.
    c) Solve the equation, and hence find the numbers that Hajra and Chris thought of.

4 Two squares are drawn. One square has sides x cm long. The other square has sides 4 cm longer. The total area of **both** the squares is 208 cm².
   a) Write down an equation in x.
   b) Show that this equation can be expressed as $x^2 + 4x - 96 = 0$.
   c) Solve the equation to find the value of x.

5 The area of a rectangle is 40 cm².
   One side of the rectangle measures x cm and the other measures $(x + 3)$ cm.
   a) Express this information as an equation in x.
   b) Show that this equation can be expressed as $x^2 + 3x - 40 = 0$.
   c) Solve your equation, and hence find the dimensions of the rectangle.

6 A rectangle measures x cm by $(x + 5)$ cm.
   A second rectangle measures $(2x - 5)$ cm by $(x + 8)$ cm.
   a) Write down expressions for the areas of the two rectangles.
   Both rectangles have the same area.
   b) Show that $x^2 + 6x - 40 = 0$.
   c) Solve this equation.
   d) Hence determine the dimensions of each rectangle.

EXERCISE 25.4 pt

1 Write the following expressions in the form $(x + a)^2 + b$, giving the values of a and b:
   a) $x^2 + 6x + 10$ **b)** $x^2 + 2x + 5$

2 Write the expression $x^2 - 4x + 8$ in the form $(x + p)^2 + q$, giving the values of p and q.

3 Write the expression $x^2 - 10x - 15$ in the form $(x + f)^2 + g$, giving the values of f and g.

4 Write the expression $x^2 + 8x + 20$ in the form $(x + a)^2 + b$, giving the values of a and b.

5 a) Write the expression $x^2 - 20x + 10$ in the form $(x + a)^2 + b$, giving the values of a and b.
   b) Sketch the graph of $y = x^2 - 20x + 10$, marking the coordinates of the minimum point on the curve.

6 a) Write the expression $x^2 + 12x + 5$ in the form $(x + a)^2 + b$, giving the values of a and b.
   b) Sketch the graph of $y = x^2 + 12x + 5$, marking the coordinates of the minimum point on the curve.

7 a) Write the expression $x^2 - 2x - 7$ in the form $(x + a)^2 + b$, giving the values of a and b.
   b) Sketch the graph of $y = x^2 - 2x - 7$, marking the coordinates of the minimum point on the curve.

8 a) Write the expression $x^2 + 14x + 100$ in the form $(x + a)^2 + b$, giving the values of a and b.
   b) Sketch the graph of $y = x^2 + 14x + 100$, marking the coordinates of the minimum point on the curve.

9 a) Use the method of completing the square to help you sketch the graph of the function $y = x^2 + 30x + 20$.
   b) The graph has a vertical line of symmetry. Give the equation of this line.

25 Quadratic equations

CHAPTER 26

Advanced algebra

EXERCISE 26.1

Simplify the following surds.

1 $\sqrt{20}$ **2** $\sqrt{75}$ **3** $\sqrt{27}$ **4** $\sqrt{32}$

5 $\sqrt{28}$ **6** $\sqrt{54}$ **7** $\sqrt{500}$ **8** $\sqrt{72}$

Write each of these as a single surd term.

9 $\sqrt{50} + \sqrt{8}$ **10** $\sqrt{40} + \sqrt{90}$ **11** $\sqrt{300} - \sqrt{48}$

12 $\sqrt{80} + \sqrt{125}$ **13** $\sqrt{7} + \sqrt{63}$ **14** $\sqrt{44} - \sqrt{11}$

Simplify each of these.   **15** $\sqrt{5}(\sqrt{5} - \sqrt{7})$ **16** $(5 + \sqrt{2})(3 - \sqrt{2})$

17 $(\sqrt{6} + \sqrt{5})(\sqrt{6} - \sqrt{5})$ **18** $(5 - \sqrt{3})^2$ **19** $\dfrac{\sqrt{27} + \sqrt{12}}{\sqrt{12}}$

20 A rectangle measures $(3 + \sqrt{44})$ cm long by $(6 - \sqrt{11})$ cm wide.
    a) Write the length of the rectangle in its simplest form.
    b) Work out the perimeter of the rectangle.
        Give your answer as an exact surd, in its simplest form.
    c) Work out the area of the rectangle.
        Give your answer as an exact surd, in its simplest form.

Solve each of the following quadratic equations, using the quadratic formula.
Leave your answers in surd form.

21 $x^2 + 5x + 2 = 0$ **22** $x^2 - x - 1 = 0$ **23** $x^2 - 8x + 3 = 0$

24 $x^2 - 8x + 5 = 0$ **25** $2x^2 + 5x + 1 = 0$

EXERCISE 26.2

Express each of these as a single fraction.

1 $\dfrac{x}{2} + \dfrac{x+4}{3}$ **2** $\dfrac{x-2}{3} + \dfrac{x+5}{4}$ **3** $\dfrac{2x}{7} + \dfrac{x+3}{5}$

4 $\dfrac{x+3}{2} - \dfrac{x+1}{5}$ **5** $\dfrac{5x+3}{4} + \dfrac{x}{7}$ **6** $\dfrac{3x-1}{5} + \dfrac{x-2}{2}$

7 $\dfrac{6}{x} + \dfrac{5}{x-2}$

8 $\dfrac{1}{x-4} + \dfrac{3}{x-2}$

9 $\dfrac{5}{x-2} + \dfrac{3}{x+3}$

10 $\dfrac{4}{x+5} + \dfrac{3}{x+2}$

11 $\dfrac{7}{x+1} - \dfrac{3}{x+2}$

12 $\dfrac{3}{x+1} + \dfrac{2}{(x+1)(x+3)}$

Solve these equations involving algebraic fractions.

13 $\dfrac{x}{4} - \dfrac{x+2}{5} = 2$

14 $\dfrac{x+2}{2} + \dfrac{x-1}{5} = \dfrac{1}{10}$

15 $\dfrac{x-1}{3} = \dfrac{x+1}{4}$

16 $\dfrac{4x+1}{3} - \dfrac{3}{2} = \dfrac{2x+5}{6}$

17 $\dfrac{2x-3}{6} + \dfrac{x+2}{3} = \dfrac{5}{2}$

18 $\dfrac{3}{2x} + \dfrac{1}{x} = \dfrac{2}{5}$

19 $\dfrac{6}{x} - \dfrac{3}{2x} = 5$

20 $\dfrac{6}{x} + \dfrac{1}{x-5} = 2$

EXERCISE 26.3

Simplify these algebraic fractions.

1 $\dfrac{3x-9}{6x+3}$

2 $\dfrac{30-10x}{5x+15}$

3 $\dfrac{14}{21x+70}$

4 $\dfrac{6x^2+9x}{3x}$

5 $\dfrac{(2x-1)^7}{(2x-1)^5}$

6 $\dfrac{27(x+5)^8}{9(x+5)^2}$

7 $\dfrac{x^2-2x}{3x}$

8 $\dfrac{8x+12}{4}$

9 $\dfrac{70(5-2x)^{15}}{10(5-2x)^5}$

10 $\dfrac{x^2+4x}{x^2-3x}$

11 $\dfrac{x^2(x+1)^4}{x(x+1)^3}$

12 $\dfrac{x^2+8x}{2x^2+4x}$

Simplify fully these expressions.

13 $\dfrac{x^2+6x+5}{x+5}$

14 $\dfrac{x^2+x-12}{x^2-16}$

15 $\dfrac{x^2+2x+1}{x^2+7x+6}$

16 $\dfrac{x^2+3x+2}{x^2+10x+16}$

17 $\dfrac{x^2+8x+16}{x^2+9x+20}$

18 $\dfrac{x^2+5x-14}{x^2+4x-21}$

19 $\dfrac{x^2-25}{x^2-2x-15}$

20 $\dfrac{x^2-100}{x^2+10x}$

EXERCISE 26.4

Solve these simultaneous equations.

1 $y = x$
$y = x^2 - 20$

2 $y = x + 2$
$y = x^2 - 4$

3 $y = 4 - 5x$
$y = x^2 - 10$

4 $y = 7x + 4$
$y = 2x^2$

5 $x^2 = 3 + y$
$y = 6x - 11$

6 $y = 2x$
$y = x^2 - 15$

Solve these simultaneous equations.

7 $y = 2 - x$
$x^2 + y^2 = 20$

8 $y = 2x - 5$
$x^2 + y^2 = 25$

9 $x = 2y + 1$
$x^2 + y^2 = 13$

10 $y + 2 = 2x$
$x^2 + y^2 = 25$

Solve these simultaneous equations.

11 $x + y = 8$
$y = 10 - x^2$

12 $x + y = 10$
$x^2 + y^2 = 122$

13 $y = 5x$
$y = x^2 + 2x - 18$

14 $y = 2x^2 - 7$
$y = 2x + 5$

15 $y = 2x + 1$
$x^2 + y^2 = 13$

16 $y = 3 - x$
$x^2 + y^2 = 29$

EXERCISE 26.5

Make x the subject of these equations.

1 $xr - t = px$

2 $px + q = rx + 5$

3 $5(x + k) = 3xy$

4 Make c the subject of the equation: $y(x + c) = 2(x - c)$

5 Make t the subject of the equation: $\dfrac{5 + t}{t - r} = w$

Make x the subject of these equations.

6 $2x + a = cx - d$

7 $\dfrac{a}{x} + 2 = c$

8 $\dfrac{xy - 5}{x} = 1$

9 $y = \dfrac{2x + k}{x - 2}$

10 Make e the subject of the equation: $\dfrac{d}{4} - \dfrac{e}{5} = 3$

EXERCISE 26.6

1 The graph shows the value of a motorbike at the end of each year.
The value of the motorbike y at time t is given by the relation:

$$y = p \times q^t$$

The curve passes through the point $(0, 1500)$.
a) Use this information to show that $p = 1500$.
The curve also passes through $(3, 444)$.
b) Use this information to find the value of q.
c) Work out the value of the motorbike after 4 years.

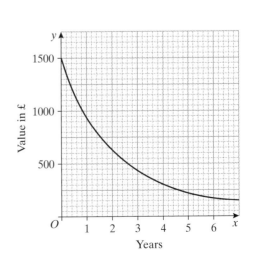

2 Ivan buys a car for £16 000.
Each year it falls in value by 40% of its value at the beginning of that year.
Ivan uses a mathematical equation to describe the value £V of his car y years after he bought it.
His equation is of the form $V = p \times q^y$, where p is a whole number and q is a decimal less than 1.
a) Write down the values of p and q.
b) Use Ivan's equation, with your values from part a), to find the value of his car after 7 years.

3 Betty's Better Biscuit company employed 1200 people in 1992.
They had to reduce the number of workers by 8% per year for each year after 1992.
a) Write down the number of employees in
   (i) 1993   (ii) 1994.
   Write your answer to the nearest whole number.
b) The number of workers n years after 1992 is given by an expression of the form $p \times q^n$.
   Write down the values of p and q.
c) Hence find the number of employees in 2001.
   Write your answer to the nearest whole number.

4 Jasmine bought a house.
Each year after that, its value increased by a fixed percentage of its value at the start of that year.
When the house was n years old its value was £130 000 $\times$ 1.04^n.
a) Write down how much Jasmine paid for the house.
b) Write down the percentage increase in value each year.
c) Work out the value of the house when it was 25 years old.
   Write your answer correct to the nearest £1000.

5 The number of ants, N, in a colony is given by the formula $N = 1600 + 850 \times 1.2^t$.
where t is the number of months after the total was first counted.
a) Find the number of ants in the colony when it was first counted.
b) Work out the number of ants in the colony 6 months after it was first counted.
c) Work out how many months it took after the first count for the colony to have over 10 000 ants.

CHAPTER 27

Further trigonometry

EXERCISE 27.1

Find the lengths of the sides represented by letters.
Give your answers to 3 significant figures.

1

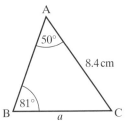

2

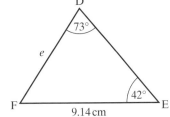

3

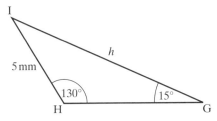

4

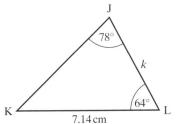

5

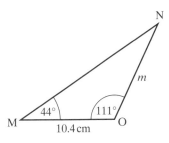

6

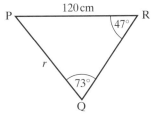

7

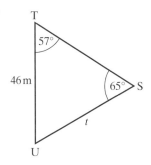

8

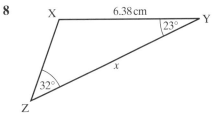

Find the angles represented by *x*. Give your answers to the nearest 0.1°.

9

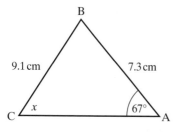

B
9.1 cm   7.3 cm
C   *x*   67°   A

10

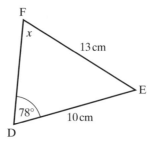

F
x   13 cm
78°   E
D   10 cm

11

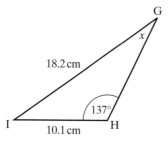

G
x
18.2 cm
137°
I   10.1 cm   H

12

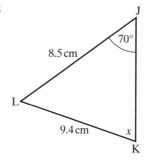

J
70°
8.5 cm
L
9.4 cm   *x*
K

EXERCISE 27.2 pt

Find the angles represented by letters. Give your answers to the nearest 0.1°.
If any are ambiguous, give both possibilities.

★ **1**

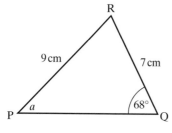

R
9 cm   7 cm
68°
P   *a*   Q

2

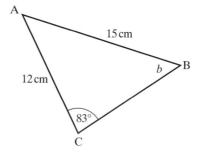

A
15 cm
12 cm   *b*   B
83°
C

3

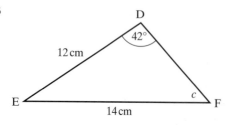

D
42°
12 cm
c
E   F
14 cm

4

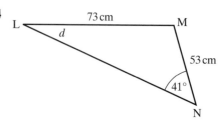

L   73 cm   M
d
53 cm
41°
N

(contd.)

5

P
114°
11 cm
e Q
20 cm
R

6

X
10 cm
7 cm
85°
Y *f* Z

7

G
9.23 cm
12.4 cm
59° F
g
E

8

4.7 cm
H 146° I
10.3 cm
h
J

Use the cosine rule to find the unknown sides indicated by letters in these triangles.

1

P
a
4 cm
62°
Q 7 cm
R

2

F
b
8 cm
110°
D 6 cm E

3

Y
c
9 cm
X
5 cm 42°
Z

4

A
d
12 cm
132°
B 15 cm
C

5

F
72°
42 m
50 m
E
e
G

6

K
8 cm
f
40°
H 6 cm J

27 **Further trigonometry**

Use the cosine rule to find the unknown angles indicated by letters in these triangles.

7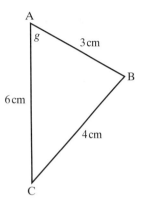

8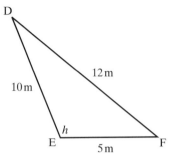

9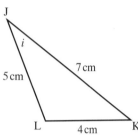

10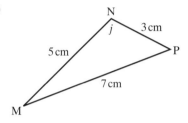

 EXERCISE 27.3B

These are mixed questions on the sine and cosine rules.
Find the missing sides represented by letters.
Give your answers correct to 3 significant figures.

1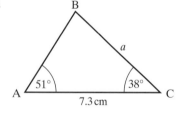

2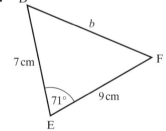

3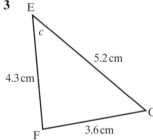

4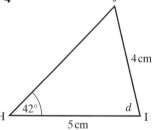

5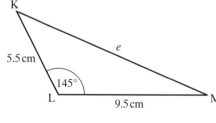

6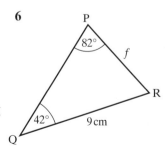

Find the unknown angles indicated by letters. Give your answers to the nearest 0.1°.

7

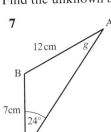

8

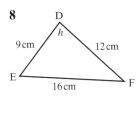

9

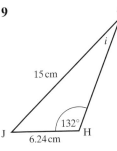

10

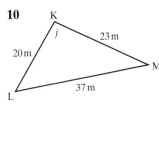

EXERCISE 27.4

1 In triangle PQR, RQ = 6.4 cm, angle PRQ = 43°
and angle RPQ = 71°. Calculate:
  a) PR
  b) the area of triangle PQR.
    Give your answer correct to 3 significant figures.

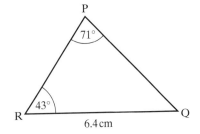

2 In triangle PQR, PQ = 13.7 cm, QR = 19.3 cm and
PR = 12.1 cm.
Calculate
  a) the value of x
  b) the area of triangle PQR.
    Give your answer correct to 3 significant figures.

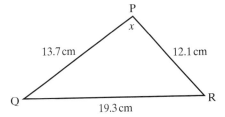

3 In triangle ABC, AB = 10 cm and BC = 5 cm.
Angle ABC = 110°.
  a) Calculate AC.
    Give your answer correct to 3 significant figures.
  b) Calculate the size of angle BAC.
    Give your answer correct to the nearest 0.1°.
  c) Calculate the size of angle ACB.
    Give your answer correct to the nearest 0.1°.
  d) Calculate the area of triangle ABC.
    Give your answer correct to 3 significant figures.

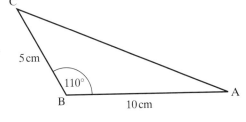

4 In triangle PQR, PQ = 5 cm, QR = 8 cm
and RP = 12 cm.
  a) Calculate the size of angle PQR.
    Give your answer correct to the nearest 0.1°.
  b) Calculate the area of triangle PQR.
    Give your answer correct to 3 significant figures.

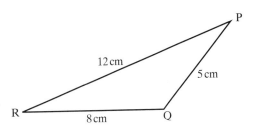

5 A chord PQ is drawn across a circle of radius 7 cm.
The chord PQ is of length 10 cm.
   a) Use the cosine rule to find angle POQ.
   Give your answer to the nearest 0.1°.
   b) Hence find the area of the segment shaded in
   the diagram.

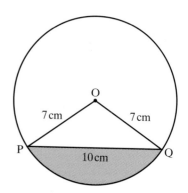

6 In triangle ABC, AB = 12 cm and AC = 10 cm.
The area of triangle ABC is 40 cm².
   a) Calculate the size of angle BAC.
   Give your answer correct to the nearest 0.1°.
   b) Calculate the perimeter of triangle ABC.
   Give your answer correct to 3 significant figures.

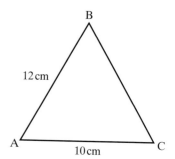

7 The diagram shows a regular pentagon inscribed inside a circle of radius 8 cm.

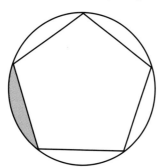

   a) Work out the area of the circle.
   b) Work out the area of the segment shaded on the diagram.
   c) Hence work out the area of the pentagon.
   Give all your answers correct to 3 significant figures.

8 An isosceles triangle has sides of length 5 cm, 5 cm and 7 cm.
Work out the area of this triangle.

.7 Further trigonometry

EXERCISE 27.5 pt

1 The diagram shows a box in the shape of a cuboid ABCDEFGH.

    AB = 5 cm, BC = 4 cm, CG = 3 cm.

A string runs diagonally across the box from A to G.
  a) Calculate the length of the string AG.
  Give your answer correct to 3 significant figures.
  b) Work out the angle between the string AG and the horizontal plane ABCD.
  Give your answer correct to the nearest 0.1°.

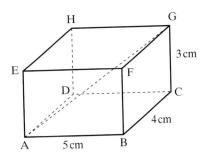

2 The diagram shows a square-based pyramid ABCDX.
AB = BC = 8 cm.
The point M is the centre of the square base ABCD.
XM = 10 cm.
  a) Calculate the length of AC.
  Give your answer correct to 3 significant figures.
  b) Work out the angle between the edge AX and the horizontal plane ABCD.
  Give your answer correct to the nearest 0.1°.

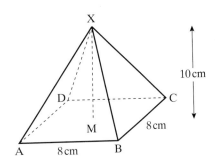

3 The diagram shows a wedge in the shape of a prism PQRSUV.

    PQ = 9 cm, QR = 18 cm, UR = 15 cm.

String runs diagonally across the box from P to U.
  a) Calculate the angle UQR.
  Give your answer correct to the nearest 0.1°.
  b) Calculate the length PU.
  Give your answer correct to 3 significant figures.
  c) Work out the angle between PU and the horizontal plane PQRS.
  Give your answer correct to the nearest 0.1°.

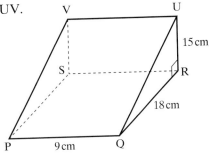

4 The diagram shows a cuboid ABCDEFGH.
AB = 7 cm, BC = 10 cm, AG = 15 cm.
  a) Calculate the length AE.
  Give your answer correct to 3 significant figures.
  b) Work out the angle between CE and the horizontal plane ABCD.
  Give your answer correct to the nearest 0.1°.

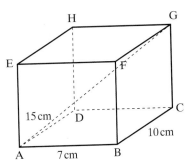

5 The cuboid PQRSJKLM has a square base PQRS.
PQ = QR = 12 cm, LR = 9 cm.

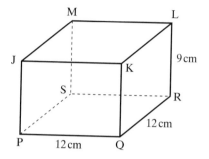

a) Calculate the length QS.
Give your answer correct to 3 significant figures.

b) Calculate the length QM.
Give your answer correct to 3 significant figures.

c) Work out the angle between QM and the horizontal plane PQRS.
Give your answer correct to the nearest 0.1°.

EXERCISE 27.6

1 A cone of height 12 cm has a base radius of 9 cm.

a) Work out the volume of the cone.

A smaller cone of height 5 cm is removed from the top of this cone to make a frustum.

b) Work out the base radius of the smaller cone.

c) Work out the volume of the frustum.

2 The diagram shows a cone of height 10 cm.
The slant height of the cone, indicated on the diagram, is 14 cm.

a) Work out the radius of the base of the cone.

b) Hence work out the volume of the cone.

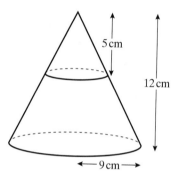

The cone is cut by removing a smaller cone of height 5 cm, to leave a frustum.

c) Work out the volume of the frustum.

3 Find the volume of this frustum of a cone.
Give your answer correct to 3 significant figures.

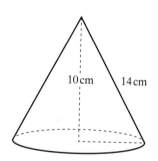

CHAPTER 28

Graphs of curves

EXERCISE 28.1

1 Complete the table of values for the function $y = x^2 + 7$.
Two values have been filled in for you.

x	-3	-2	-1	0	1	2	3
y		11					16

2 Complete the table of values for the function $y = 3x^2 + x$.
Two values have been filled in for you.

x	-2	-1	0	1	2	3
y		2				30

3 Complete the table of values for the function $y = x - x^2$.
Two values have been filled in.

x	-2	-1	0	1	2	3
y		-2			-2	

4 Complete the table of values for the function $y = x^2 + 2x$.

x	-3	-2	-1	0	1	2
y						

5 Complete the table of values for the function $y = x^3 - 3$.

x	-2	-1	0	1	2	3
y						

6 Complete the table of values for the function $y = \dfrac{6}{x}$.

x	-3	-2	-1	0	1	2	3
y				not defined			

7 Complete the table of values for the function $y = 3^x$.

x	-2	-1	0	1	2	3
y						

8 Complete the table of values for the function $y = x + 2x^2$.

x	-2	-1	0	1	2	3	4
y							

EXERCISE 28.2

1 The table shows values of the function $y = x^2 - 2$ for $-3 \leqslant x \leqslant 3$.

x	-3	-2	-1	0	1	2	3
y	7	2	-1	-2	-1	2	7

a) Draw a set of coordinate axes on squared paper, so that x can range from -3 to 3 and y from -3 to 8, and plot these points on your axes.
b) Join your points with a smooth curve.
c) Use your graph to find the value of y when $x = 2.5$.
d) Find the coordinates of the lowest point on the curve.

2 The table shows some values of the function $y = 3x - x^2$.

x	-2	-1	0	1	2	3	4	5
y		-4			2			-10

a) Copy and complete the table.
b) Draw a set of coordinate axes on squared paper, so that x can run from -2 to 5 and y from -10 to 4. Plot these points on your graph, and join them with a smooth curve.
c) Give the coordinates of the highest point on the graph.

3 The table shows some values of the function $y = 3x^2 - 5$.

x	-3	-2	-1	0	1	2	3
y		7			-2		

a) Copy and complete the table.
b) Draw a set of coordinate axes on squared paper, so that x can run from -3 to 3 and y from -6 to 25. Plot these points on your graph, and join them with a smooth curve.
c) Use your graph to find all the solutions to the equation $3x^2 - 5 = 0$.

4 The table shows some values for the function $y = x^3 + 15$.

x	-3	-2	-1	0	1	2	3
y	-12				16		

a) Copy and complete the table.
b) Draw a set of coordinate axes on squared paper, so that x can run from -3 to 3 and y from -15 to 50. Plot these points on your graph, and join them with a smooth curve.

c) Use your graph to solve the equation $x^3 + 15 = 0$.
d) Use your graph to solve the equation $x^3 + 15 = 20$.

5 The table shows some values of the function $y = 2^x - 3$.

x	−2	−1	0	1	2	3	4
y	−2.75		−2				13

a) Copy and complete the table.
b) Draw a set of coordinate axes on squared paper, so that x can run from −2 to 4 and y from −3 to 14. Plot these points on your graph, and join them with a smooth curve.
c) Use your graph to solve the equation $2^x - 3 = 11$.

6 The table shows some values of the function $y = 2 - x^2$.

x	−3	−2	−1	0	1	2	3
y		−2		2			−7

a) Copy and complete the table.
b) Draw a set of coordinate axes on squared paper, so that x can run from −3 to 3 and y from −8 to 3. Plot these points on your graph, and join them with a smooth curve.
c) State the coordinates of the point on the curve where y takes its maximum value.
d) Use your graph to find the two solutions to the equation $2 - x^2 = 0$.

7 The table shows some values of the function $y = \dfrac{8}{x}$. The function is not defined when $x = 0$.

x	−8	−4	−2	−1	0	1	2	4	8
y			−4		not defined		4		

a) Copy and complete the table.
b) Draw a set of coordinate axes on squared paper, so that x can run from −8 to 8 and y from −8 to 8. Plot these points on your graph.
c) Join the first four points with a smooth curve.
d) Join the last four points with another smooth curve.
e) Use your graph to solve the equation $\dfrac{8}{x} = 5$.

8 The table shows some x values for the function $y = 2x^2 + x - 6$.

x	−3	−2	−1	0	1	2	3
y							

a) Copy and complete the table.
b) Draw a set of coordinate axes on squared paper, so that x can run from −3 to 3 and y from −7 to 16. Plot these points on your graph, and join them with a smooth curve.
c) Write down the solutions to the equation $2x^2 + x - 6 = 0$.
d) Give the coordinates of the minimum point on the curve.

9 The diagram shows part of the graph of the
function $y = x^2 - 4x + 2$.
   a) Use the graph to find the two solutions
   to the equation $x^2 - 4x + 2 = 0$.
   b) Use the graph to find the two solutions
   to the equation $x^2 - 4x + 2 = 1$.

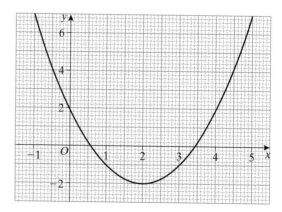

10 The diagram shows part of the graph of $y = 2 + \dfrac{1}{x}$.

Use the graph to find a solution to the equation $2 + \dfrac{1}{x} = 4.5$.

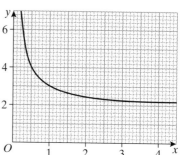

EXERCISE 28.3

1 Draw three sketch graphs to show
   a) $y = \sin x$ **b)** $y = \cos x$ **c)** $y = \tan x$
   as x ranges from $-90°$ to $360°$.

2 Here are some statements about trigonometric functions. Decide whether each one is true or
false. Try to deduce the answer from graphical considerations rather than using a calculator!
   a) $\sin 210°$ and $\sin 330°$ have the same value.
   b) $\cos 210°$ and $\cos 330°$ have the same value.
   c) $\sin x$ always lies between 0 and 1.
   d) $\tan x$ cannot be negative provided x lies between 0 and $180°$.
   e) $\tan 30°$ and $\tan 210°$ have the same value.

3 Here are some clues about trigonometric functions.
   For each one, decide whether it is referring to $y = \sin x$, $y = \cos x$ or $y = \tan x$.
   a) This function has a minimum when $x = 270°$.
   b) This function is not defined when $x = 270°$.
   c) The graph of this function passes through the point $(180°, 0)$.
   d) The graphs of these two functions have rotational symmetry order 2.
   e) This function has a minimum when $x = 180°$.

✗ EXERCISE 28.4

1 The diagram shows part of
the graph of $y = f(x)$.

On a copy of the diagram,
draw the graphs of:
a) $y = f(x - 2)$
b) $y = f(x) - 3$
c) $y = 2f(x)$
Label your three graphs
clearly.

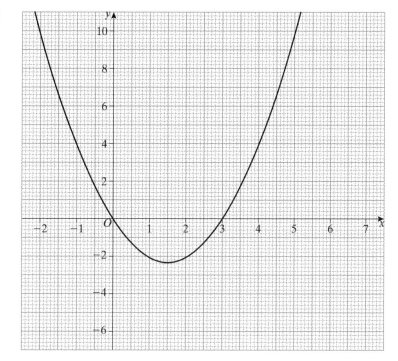

2 The diagram shows part of the graph of $y = f(x)$.

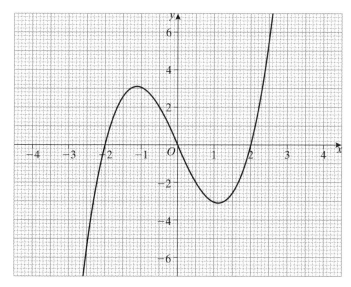

On a copy of the diagram, draw the graphs of:
a) $y = f(x + 1)$ **b)** $y = f(x) + 2$ **c)** $y = -f(x)$
Label your three graphs clearly.

3 The diagram shows part of the graph of $y = f(x)$.

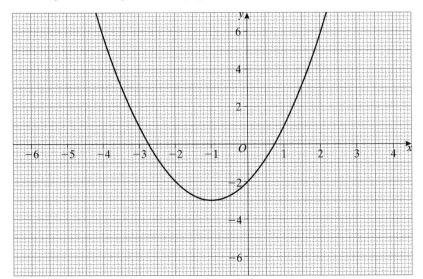

On a copy of the diagram, draw the graphs of:
a) $y = f(x - 2)$ **b)** $y = f(x - 2) - 3$ **c)** $y = -f(x)$
Label your three graphs clearly.

4 The diagram shows part of the graph of $y = f(x)$.
On a copy of the diagram, draw the graphs of:
a) $y = f(-x)$
b) $y = -f(x)$
c) $y = f(x) + 2$
Label your three graphs clearly.

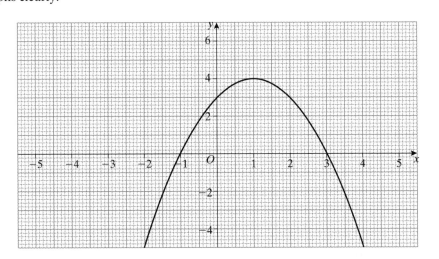

5 The diagram shows part of the graph of $y = a\cos(x - b)$, where a and b are constants. Work out the values of a and b. Explain your reasoning.

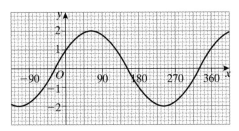

28 Graphs of curves

6 The diagram shows part of the graphs of $y = f(x)$ and $y = f(x + a) + b$.

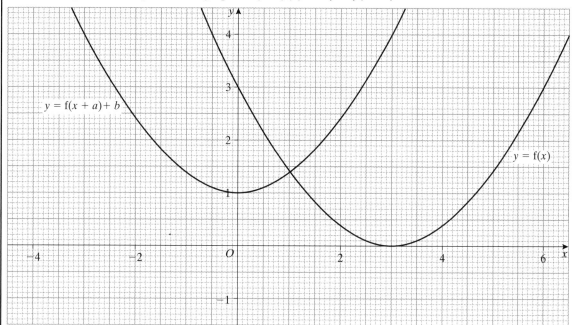

Work out the values of a and b. Explain your reasoning.

7 The diagram shows part of the graph of $y = a\sin(bx)$.

Work out the values of a and b. Explain your reasoning.

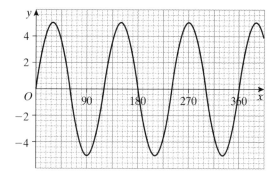

8 The diagram shows part of the graph of $y = a\sin x$.

Work out the value of a. Explain your reasoning.

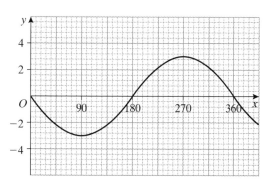

CHAPTER 29

Vectors

EXERCISE 29.1

1 The diagram below shows some vectors drawn on a grid of unit squares.
Write down column vectors to describe each one.

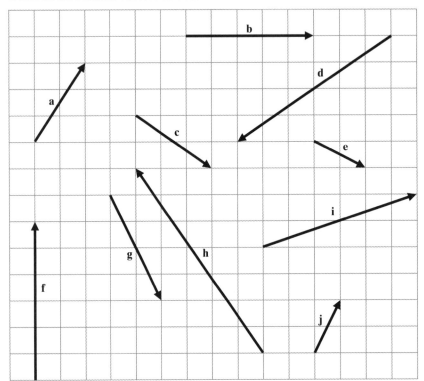

2 Draw a sketch of each of these vectors on squared paper.

a) $\begin{bmatrix} 4 \\ 2 \end{bmatrix}$
b) $\begin{bmatrix} -3 \\ 5 \end{bmatrix}$
c) $\begin{bmatrix} 6 \\ -1 \end{bmatrix}$
d) $\begin{bmatrix} -2 \\ -3 \end{bmatrix}$

EXERCISE 29.2

The vectors **p**, **q** and **r** are given by $\mathbf{p} = \begin{bmatrix} 6 \\ 3 \end{bmatrix}$, $\mathbf{q} = \begin{bmatrix} -1 \\ -4 \end{bmatrix}$ and $\mathbf{r} = \begin{bmatrix} -5 \\ 2 \end{bmatrix}$.

Work out each of these as a column vector. Illustrate your answer with a diagram.

1 **p** + **q**
2 **r** − **p**
3 **q** + **r**
4 **q** − **p**

Work out:

5 $\mathbf{p} - \mathbf{r} + \mathbf{q}$ **6** $\mathbf{r} - \mathbf{q} + \mathbf{p}$

The vectors $\mathbf{c}$, $\mathbf{d}$ and $\mathbf{e}$ are given by $\mathbf{c} = \begin{bmatrix} 7 \\ -3 \end{bmatrix}$, $\mathbf{d} = \begin{bmatrix} 4 \\ 5 \end{bmatrix}$ and $\mathbf{e} = \begin{bmatrix} -2 \\ 1 \end{bmatrix}$.

Work out each of these as a column vector. Illustrate your answer with a diagram.

7 $\mathbf{e} - \mathbf{d}$ **8** $\mathbf{c} + \mathbf{d}$ **9** $\mathbf{d} - \mathbf{c}$ **10** $\mathbf{c} + \mathbf{d} + \mathbf{e}$

Work out:

11 $\mathbf{c} - \mathbf{e}$ **12** $\mathbf{e} - \mathbf{d} - \mathbf{c}$

13 You are given that $\begin{bmatrix} 5 \\ 3 \end{bmatrix} + \begin{bmatrix} x \\ -1 \end{bmatrix} = \begin{bmatrix} 3 \\ 2 \end{bmatrix}$. Find the value of x.

14 You are given that $\begin{bmatrix} 4 \\ x \end{bmatrix} - \begin{bmatrix} y \\ -7 \end{bmatrix} = \begin{bmatrix} 7 \\ 9 \end{bmatrix}$. Find the values of x and y.

15 You are given that $\begin{bmatrix} -5 \\ -8 \end{bmatrix} + \begin{bmatrix} 8 \\ y \end{bmatrix} = \begin{bmatrix} x \\ 2 \end{bmatrix}$. Find the values of x and y.

EXERCISE 29.3

The vectors $\mathbf{x}$, $\mathbf{y}$ and $\mathbf{z}$ are given by $\mathbf{x} = \begin{bmatrix} 8 \\ 2 \end{bmatrix}$, $\mathbf{y} = \begin{bmatrix} -4 \\ -1 \end{bmatrix}$ and $\mathbf{z} = \begin{bmatrix} -5 \\ 7 \end{bmatrix}$.

Work out:

1 $2\mathbf{x}$ **2** $\mathbf{z} + 3\mathbf{y}$ **3** $2\mathbf{y} - \mathbf{z}$

4 $10\mathbf{x} - 5\mathbf{z}$ **5** $\mathbf{y} + 3\mathbf{x} - 2\mathbf{z}$ **6** $2\mathbf{x} - 3\mathbf{y} + \mathbf{z}$

The vectors $\mathbf{d}$, $\mathbf{e}$ and $\mathbf{f}$ are given by $\mathbf{d} = \begin{bmatrix} 6 \\ 0 \end{bmatrix}$, $\mathbf{e} = \begin{bmatrix} 3 \\ -5 \end{bmatrix}$ and $\mathbf{f} = \begin{bmatrix} -4 \\ 1 \end{bmatrix}$.

Work out:

7 $2\mathbf{d}$ **8** $-2\mathbf{e}$ **9** $3\mathbf{f} - \mathbf{d}$

10 $5\mathbf{d} - \mathbf{e} + 4\mathbf{f}$ **11** $8\mathbf{e} - 6\mathbf{f}$ **12** $4\mathbf{d} - 3\mathbf{f} + 2\mathbf{e}$

13 You are given that $4\begin{bmatrix} 3 \\ 9 \end{bmatrix} + 2\begin{bmatrix} x \\ -1 \end{bmatrix} = \begin{bmatrix} 22 \\ 34 \end{bmatrix}$. Find the value of x.

14 You are given that $3\begin{bmatrix} x \\ y \end{bmatrix} - 5\begin{bmatrix} 0 \\ -2 \end{bmatrix} = \begin{bmatrix} 12 \\ 16 \end{bmatrix}$. Find the values of x and y.

15 You are given that $6\begin{bmatrix} 5 \\ -2 \end{bmatrix} + 3\begin{bmatrix} 8 \\ x \end{bmatrix} = \begin{bmatrix} y \\ -21 \end{bmatrix}$. Find the values of x and y.

X EXERCISE 29.4 pt

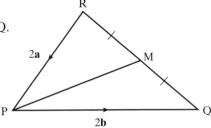

1 The diagram shows triangle RPQ. M is the midpoint of RQ.

$\overrightarrow{RP} = 2\mathbf{a}$ and $\overrightarrow{PQ} = 2\mathbf{b}$.

Find in terms of **a** and **b**, expressions for

a) $\overrightarrow{RQ}$ b) $\overrightarrow{RM}$

c) $\overrightarrow{PM}$ d) $\overrightarrow{MP}$

2 The diagram shows a parallelogram ABCD.
AD is produced to E.

$\overrightarrow{DE} = \mathbf{p}$ and $\overrightarrow{CE} = \mathbf{q}$. $DE = \frac{1}{3}AD$.

Find, in terms of **p** and **q**, expressions for:

a) $\overrightarrow{AD}$ b) $\overrightarrow{AB}$ c) $\overrightarrow{BD}$

d) What kind of quadrilateral is BCED?

3 A quadrilateral ABCD is made by joining the points A $(0, 3)$, B $(3, 1)$, C $(5,4)$ and D $(2, 6)$.

a) Write down column vectors for

    (i) $\overrightarrow{AB}$ (ii) $\overrightarrow{DC}$

b) What do your answers tell you about AB and DC?

c) Write down column vectors for

    (i) $\overrightarrow{BC}$ (ii) $\overrightarrow{AD}$

d) What kind of quadrilateral is ABCD?

4 The diagram shows a quadrilateral ABCD. M is the midpoint of AC.

$\overrightarrow{AB} = 2\mathbf{p}$, $\overrightarrow{BM} = \mathbf{q}$ and $\overrightarrow{AD} = 3\mathbf{p} + 2\mathbf{q}$.

Find, in terms of **p** and **q**, expressions for:

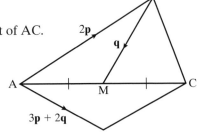

a) $\overrightarrow{AM}$ b) $\overrightarrow{AC}$

c) $\overrightarrow{CB}$ d) $\overrightarrow{DC}$

e) What kind of quadrilateral is ABCD?

5 The diagram shows a parallelogram PQRS.

$\overrightarrow{PQ} = 2\mathbf{a}$ and $\overrightarrow{PS} = 2\mathbf{b}$.

L is the midpoint of PR.
M is the midpoint of QS.
Find, in terms of **a** and **b** expressions for:

a) $\overrightarrow{SR}$ b) $\overrightarrow{PR}$

c) $\overrightarrow{LM}$ d) $\overrightarrow{QM}$

e) Write down two facts about the relationship between LM and PQ.

6 A quadrilateral ABCD is made by joining A $(1, 3)$, B $(1, -5)$, C $(-2, -5)$ and D $(-2, 3)$.

   a) Write column vectors for:

      (i) $\overrightarrow{AB}$ **(ii)** $\overrightarrow{DC}$

   b) What do your answers to part a) tell you about AB and DC?

   c) Write column vectors for:

      (i) $\overrightarrow{CB}$ **(ii)** $\overrightarrow{DA}$

   d) What kind of quadrilateral is ABCD?

7 The diagram shows a kite ABCD. $\overrightarrow{AB} = 2\mathbf{a}$, $\overrightarrow{AD} = 2\mathbf{b}$ and $\overrightarrow{CD} = 2\mathbf{c}$.
E, F, G and H are the midpoints of AB, BC, CD and DA respectively.

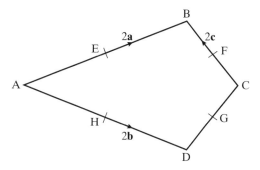

   a) Explain why $\overrightarrow{DC} = 2\mathbf{a} - 2\mathbf{b} - 2\mathbf{c}$.

   b) Find in terms of **a**, **b** and **c**:

      (i) $\overrightarrow{EF}$ **(ii)** $\overrightarrow{HG}$

      (iii) $\overrightarrow{EH}$ **(iv)** $\overrightarrow{FG}$

   c) What can you deduce about the line segments EH and FG?

   d) What type of quadrilateral is EFGH?

CHAPTER 30

Mathematical proof

 EXERCISE 30.1

1 Look at these pairs of triangles. In each case, decide whether the triangles are congruent or not. If they are congruent, explain why.

a)

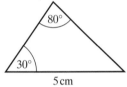

b)

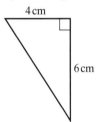

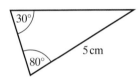

c)

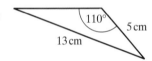

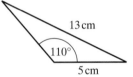

d)

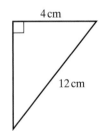

e)

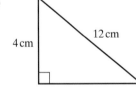

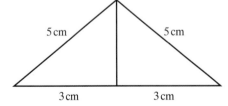

f)

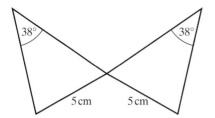

g)

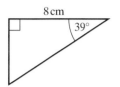

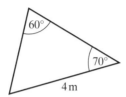

h)

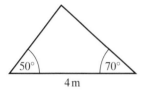

2 The diagram shows two tangents, AB
and AC, to a circle, centre O.
   a) Show that the line segment OA bisects
   angle BOC.
   [Hint: join OA and consider triangles
   AOB and AOC.]
   b) Prove that OABC is a cyclic quadrilateral.

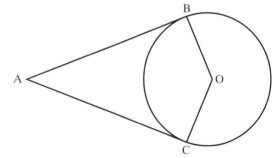

3 The diagram shows two triangles PQS and QRT.
   Q bisects the line PR and QS = RT. QS is parallel to RT.

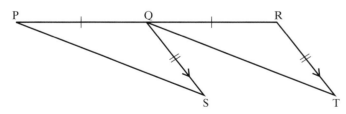

Prove that PS = QT.

4 The diagram shows a circle, centre O.
AB is a chord.
OM is perpendicular to AB.
Prove that M bisects AB.

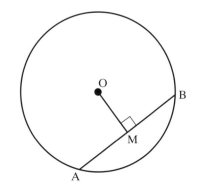

5 Q is the midpoint of the line segment PR and the
line segment ST.
Prove that triangles PQS and RQT are congruent.

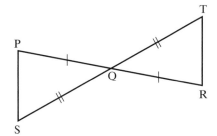

X EXERCISE 30.2

1 Prove that the sum of any two consecutive odd numbers is always a multiple of 4.

2 Prove that the sum of any five consecutive integers is always a multiple of 5.

3 Prove that the product of any two consecutive even numbers is always even.

4 Prove that the difference between any two consecutive integers is always 1.

5 Prove that the sum of the squares of any two consecutive numbers is always odd.

6 Look at this number pattern.

$$2^2 - 1^2 = 4 - 1 = 3 = 2 + 1$$
$$3^2 - 2^2 = 9 - 4 = 5 = 3 + 2$$
$$4^2 - 3^2 = 16 - 9 = 7 = 4 + 3$$
$$5^2 - 4^2 = 25 - 16 = 9 = 5 + 4$$

   a) Write down the next two lines in this number pattern.
   b) Show that the difference between the squares of any two consecutive numbers is equal to
      the sum of these two consecutive numbers by looking at:
      (i) $(n)^2 - (n - 1)^2$ **(ii)** $x^2 - y^2$

7 a) Factorise $31^2 - 29^2$.   **b)** Hence show that $31^2 - 29^2 = 120$.

8 Show that the difference between the squares of any two consecutive even numbers is a multiple
of 4.

EXERCISE 30.3

1 Anton says, 'The difference between any two prime numbers will always be a multiple of 2.'
Show that Anton is wrong by giving two counter-examples.

2 Bob says, 'If a quadrilateral has two pairs of opposite sides equal in length then it must be a rectangle.' Show that this statement is false.

3 Carla says, 'If you multiply any two factors of 48 that are 8 or less, you will always get a factor of 48.' Show that Carla's statement is wrong.

4 Don says, 'You can find the lowest common denominator when adding two fractions less than one just by multiplying the two denominators together.'
Show that this statement is false.

5 Evan says, 'If the four sides of a quadrilateral all have the same length, then the four internal angles must all be equal.'
Draw a diagram to show that Evan is wrong.

6 The perfect squares are 1, 4, 9... which are numbers of the form n^2 where n is an integer.
Integers to the power of 5 are 1, 32, 243... which are numbers of the form n^5 where n is an integer.
David says, 'Apart from 1, there is no other number which is a perfect square and also a number of the form n^5.'
Find a counter-example to show that David is wrong.

7 'If x is positive, then $4 + 4x - x^2$ is also positive.' Show that this statement is false.

8 Hans says, 'If you square a number the answer is always bigger than that number.'
Show that Hans is wrong.

9 Ian says, 'If you have any quadrilateral that has the opposite sides equal in length, there will always be at least one line of reflection symmetry.'
Draw a counter-example to show that Ian is wrong.

10 Joe says 'If n is a positive integer, then the value of $n^2 + n + 35$ is always prime.'
Show that Joe is wrong.

CHAPTER 31

Introducing coordinate geometry

 EXERCISE 31.1

1 Use Pythagoras' theorem to calculate the distance from:
   a) A $(6, 2)$ to B $(3, 7)$ **b)** P $(-2, -7)$ to Q $(3, 5)$ **c)** M $(-1, 4)$ to N $(5, -1)$.

2 A triangle ABC has vertices A $(-3, -1)$, B $(1, 1)$ and C $(-3, 4)$.
   Daisy makes a sketch of the triangle. Daisy says that the triangle is isosceles.
   a) Use Pythagoras' theorem to find the lengths of AB, AC and BC.
      You can give answers in surd form.
   b) Use your answers to decide whether Daisy is right or wrong.

3 A quadrilateral PQRS has vertices P $(0, 1)$, Q $(3, 2)$, R $(5, 6)$ and S $(2, 5)$.
   a) Make a rough sketch of the quadrilateral.
   b) Use Pythagoras' theorem to find the length of the sides PQ, QR, RS and SP.
      Give exact answers in surd form.
   c) What type of quadrilateral is PQRS?

4 Three points have coordinates A $(1, 5, 3)$, B $(6, 1, -5)$ and C $(-2, -1, 4)$.
   Use the 3-D form of Pythagoras' theorem to find, as exact surds, the lengths of:
   a) AB   **b)** BC   **c)** CA.
   d) Which side of triangle ABC is the longest?

 EXERCISE 31.2

1 A circle, centre O, has equation $x^2 + y^2 = 49$.   **2** A circle, centre O, has radius 8.
   Write down the radius of the circle.   Write down the equation of the circle.

3 A circle, centre O, has equation $x^2 + y^2 = 100$.   **4** A circle, centre O, has diameter 12.
   Work out the diameter of the circle.   Write down the equation of the circle.

5 A circle, centre O, passes through the point $(3, 7)$.   **6** A circle has equation $x^2 + y^2 = 169$.
   a) Work out the radius of the circle.   **a)** State the radius of the circle.
      Give your answer as a surd.   **b)** Find the coordinates of the two
   b) Write down the equation of the circle.   points on the circle for which $y = 5$.

7 Find the coordinates of the two points where the graph of the line $y = x + 3$ meets the circle
   with equation $x^2 + y^2 = 29$.

8 A circle C is given by the equation $x^2 + y^2 = 50$.
   A line L is given by the equation $y = x + 10$.
   a) Prove that the line L cuts the circle C in exactly one place.
   b) Find the coordinates of this point.

9 Here are the equations of three graphs.

Graph A $x = 30 + y$
Graph B $x^2 = 30 + y$
Graph C $x^2 = 30 - y^2$

a) Which one of these is a circle?
b) Describe the shapes of the other two graphs.

10 A circle C, with centre at the origin O, has diameter $\sqrt{200}$.

a) Write this diameter in the form $10\sqrt{k}$, where k is an integer to be found.
b) Hence find the equation of the circle C.

A straight line has equation $y = x + 8$.

c) Find the coordinates of the points where this line meets the circle C.

EXERCISE 31.3

1 For each of the following lines, write down:
a) The gradient of a line parallel to the given line, and
b) the gradient of a line perpendicular to the given line.

(i) $y = 4x - 1$ (ii) $y = \dfrac{x}{3}$

(iii) $y = 6 - 2x$ (iv) $2y = 9x - 4$

2 Find the equation of the straight line, parallel to $y = 2x + 7$, that passes through point P $(8, 6)$.

3 Find the equation of the straight line, parallel to $y = 3 - x$, that passes through point Q $(7, 2)$.

4 Find the equation of the straight line, perpendicular to $y = \frac{1}{2}x - 1$, that passes through the point P $(-2, -5)$.

5 Find the equation of the straight line, perpendicular to $y = 5x + 2$, that passes through the point Q $(5, 3)$.

6 Here are the equations of four straight lines.

Line A $y = 15 - 8x$ **Line B** $y = 8x - 3$ **Line C** $y - 8x + 5 = 0$ **Line D** $y = \frac{1}{8}x + 3$

a) Write down the names of the two lines that are **parallel**.
b) Write down the names of the two lines that are **perpendicular**.

7 A line L passes through the points A $(0, 7)$ and B $(3, 5)$.
a) Work out the gradient of the line L.   b) Work out the equation of the line L. Give your answer in the form $y = mx + c$.

A second line M is perpendicular to L. M passes through the point A.
c) Write down the gradient of M.   d) Work out the equation of the line M. Give your answer in the form $y = mx + c$.

8 Find the equation of the straight line, parallel to $y = 6 - 5x$, that passes through the point P $(1, 3)$.

9 Find the equation of the straight line, perpendicular to $y = 2 - x$, that passes through the point P $(-7, -4)$.

10 The lines $y = 8 - 3x$ and $y = ax + 2$ are perpendicular. Find the value of a.

CHAPTER 32

Further probability and statistics

EXERCISE 32.1

1 7 cats have a mean weight of 3.4 kg. Another cat has a weight of 4.2 kg.
Work out the mean weight of all 8 cats.

2 10 pieces of cheese have a mean weight of 320 g.
Another 8 pieces of cheese have a mean weight of 545 g.
Work out the mean weight of all 18 pieces of cheese.

3 There are 15 players in a football team.
6 of the players have a mean height of 1.7 m.
The other 9 players have a mean height of 1.62 m.
Find the mean weight for all 15 players.

4 There are 400 students at a school. The students take a Mathematics exam.
The table below shows the results.

	Number of students	Mean mark
Girls	230	63%
Boys	170	67%

Calculate the mean mark for all 400 students.

5 The mean number of children in 7 families in a street is 2.8 children.
The mean number of children in the 7 families in the next street is 1.75 children.

Leo says, 'The mean number of children in all 14 families is $\dfrac{2.8 + 1.75}{2} = 2.275$ children.'

Explain carefully whether Leo is right or wrong.

6 The mean weight loss of the 25 adults at a slimming class on Monday was 1.72 kg.
Another adult arrived late and after he was weighed the mean increased to 1.74 kg.
Work out the weight loss of the adult who arrived late.

7 To get a distinction, Katrina has to get an average of 80 marks over 10 examinations.
In the first 9 examinations Katrina obtained a mean score of 81 marks.
Work out the least number of marks she needs to score in the tenth examination in order to
obtain a distinction.

8 One year the mean monthly rainfall in a city in India was 10 cm.
The mean monthly rainfall for the first 7 months was only 5 cm.
Work out the mean monthly rainfall for the next 5 months.

EXERCISE 32.2 pt

1 The table shows the number of umbrellas sold each quarter by a small shop.
  The data was collected from January 1997 to December 1999.

	1997				1998				1999		
Jan to Mar	Apr to Jun	Jul to Sep	Oct to Dec	Jan to Mar	Apr to Jun	Jul to Sep	Oct to Dec	Jan to Mar	Apr to Jun	Jul to Sep	Oct to Dec
27	15	10	20	28	14	11	21	30	16	13	22
		18	18.25	18							

  a) Work out the four-quarter moving averages for the data.
     The first three have been done for you.
  b) Plot a time series graph of the original data.
  c) On the same axes, plot a time series graph of the moving averages.
     What do the graphs show you about the seasonal variation in the number of umbrellas sold?

2 The table shows the number of absences at a school over a period of three weeks.

	Week 1					Week 2					Week 3			
Mon	Tue	Wed	Thu	Fri	Mon	Tue	Wed	Thu	Fri	Mon	Tue	Wed	Thu	Fri
18	10	7	6	16	20	13	8	8	18	21	14	8	10	20
		11.4	11.8	12.4										

  a) Work out the five-day moving averages for the data.
     Explain why there are only 11 of these values, not 15.
     The first three have been done for you.
  b) Plot a time series graph of the original data.
     What can you say about the absences on different days of the week?
  c) Plot a time series graph of the moving averages.
     What can you say about the weekly average absences?

3 The table shows the average number of school dinners sold each day by the school canteen over
  a period of four years.

Year		2001			2002			2003			2004	
Term	1st	2nd	3rd	1st	2nd	3rd	1st	2nd	3rd	1st	2nd	3rd
	150	127	101	138	121	89	135	118	86	132	109	77
		126	122	120								

  a) Plot a time series graph of the data.
     Describe whether you can see any seasonal variation in the number of school dinners sold.
  b) Calculate a set of three-value moving averages from the data.
     Three of the ten values have been done for you.
  c) Plot a time series graph of the moving averages.
  d) Add a straight line to your graph to indicate the underlying trend.

EXERCISE 32.3

1 A drawer contains 12 socks. 3 of the socks are red and the rest are black.
Ali is going to take two socks at random from the drawer, **without** replacement.

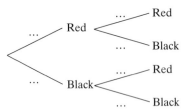

   a) Copy and complete the tree diagram.
   b) Work out the probability that Ali will take two red socks.
   c) Work out the probability that Ali will take two socks of the same colour.

2 In a class of 30 students, 9 wear glasses and 21 do not wear glasses.
Two students are chosen at random.
   a) Illustrate the situation with a tree diagram.
   b) Calculate the probability that both students wear glasses.
   c) Calculate the probability that at least one student wears glasses.

3 A small box contains 21 chocolates.
15 are milk chocolates and the rest are plain chocolates.
Two chocolates are picked at random, without replacement.
   a) Draw a tree diagram to show this information.
   b) Calculate the probability that one milk chocolate and one plain chocolate are picked.

4 Tanya forecasts the results of 3 tennis matches.
Work out the probability that she forecasts the correct result in all 3 matches.

5 A box contains 25 smarties. 8 are red, 6 are yellow and the rest are green.
Two smarties are chosen at random, without replacement.
   a) Draw a tree diagram to show this information.
   b) Calculate the probability that the two smarties are **not** the same colour.

6 In a bowl of fruit there are 3 bananas, 4 oranges and 5 apples.
Two pieces of fruit are picked at random from the bowl.
   a) Draw a tree diagram to show this information.
   b) Find the probability that 2 oranges are picked.
   c) Work out the probability that two **different** pieces of fruit are picked.

7 There are 15 students in a class.
9 of the students are right-handed, and the other 6 are left-handed.
   a) Two of the 15 students are chosen at random.
   Find the probability that they are both right-handed.
   b) Three of the 15 students are chosen at random.
   Find the probability that they are all left-handed.